Radar Development in Canada: The Radio Branch of the National Research Council of Canada 1939-1946

W. E. Knowles Middleton

This volume continues the story of the National Research Council begun by *Physics at the National Research Council of Canada* (also written by Middleton) and *Biological Sciences at the National Research Council of Canada* (by N. T. Gridgeman). Technical enough to interest the scientifically informed reader, yet comprehensible to the general reader, this history of the development of radar in Canada by the N.R.C. in the years of the Second World War explains what radar is and how it functions, and briefly describes the problems which led to the development of new equipment—such as the need to detect mortar bombs and the danger of airborne attacks on Canadian coasts. The author describes how personality clashes, tensions between co-operating organizations, and difficult administrative puzzles were overcome, allowing scientific expertise to triumph in the speedy and valuable development of new radar devices, an important contribution by Canada to the war effort.

The volume is well organized and includes illustrations. Documentation from government sources, use of quotations from correspondence and interviews, personal reminiscences of the author, and informed opinion and interpretation combine to make the volume easy and informative reading.

W. E. Knowles Middleton was associated with the National Research Council from 1946 until 1963. He is the author of The History of the Barometer, The Experimenters: A Study of the Accademia del Cimento, and Physics at the National Research Council of Canada, 1929-1952 (WLU Press, 1979). He has received honorary degrees from Boston University and McGill University.

Radar Development in Canada: The Radio Branch of the National Research Council of Canada 1939-1946

Radar Development in Canada: The Radio Branch of the National Research Council of Canada 1939-1946

W. E. Knowles Middleton

Wilfrid Laurier University Press

Canadian Cataloguing in Publication Data

Middleton, W. E. Knowles (William Edgar Knowles), 1902-
   Radar development in Canada

Includes index.
ISBN 0-88920-106-4

1. Radar – Canada – History. 2. National Research
Council of Canada. Radio and Electrical Engineering
Division – History. I. Title.

TK6576.M52   621.3848′0971   C81-094347-6

Copyright © 1981 by the National Research Council of Canada
Wilfrid Laurier University Press
Waterloo, Ontario, Canada
N2L 3C5

81 82 83 84 4 3 2 1

CONTENTS

LIST OF ILLUSTRATIONS

PREFACE

It was originally intended that the wartime history of the Radio Branch of the National Research Council of Canada should be written by someone who had been involved in the programme. For one reason or another this proved impracticable, and in the spring of 1978 I was approached by Mr. J. K. Pulfer, Director of the Electrical Engineering Division, and Dr. A. W. Tickner, Senior Archival Officer of the National Research Council, with the request that I should write this book. As I am not at all a radio engineer, I agreed to do this only because like its predecessor *Physics at the National Research Council of Canada, 1929-1952*, it is intended for the general reader. General background information may be obtained from the book by Wilfrid Eggleston, *National Research in Canada: The N.R.C., 1916-1966* (Toronto and Vancouver: Clarke, Irwin and Co., 1978).

Apart from the introductory chapters this book is confined strictly to the time between the outbreak of the Second World War and December 31, 1946. Most of the wartime projects had been completed by that time, but there were some, not the least important, that were carried on for a number of years afterwards. I have attempted no account of these further developments.

The sources will be evident from the footnotes. Dr. Tickner arranged to have an engineering student, Mrs. Barbara Chalmers, who was employed in the summer of 1978, index a large number of documents that it was necessary to consult. I also made use of a number of taped interviews obtained by Mr. D. J. C. Phillipson for the National Research Council.

For personal help I wish to thank several past and present members of the Radio and Electrical Engineering Division, especially J. E. Breeze, W. C. Brown, and J. T. Henderson. Dr. Tickner has been immensely helpful, both with administrative problems and in pointing out sources of information that I might otherwise have missed.

A draft of the manuscript was read by several members of the Electrical Engineering Division, whom I must thank for valuable suggestions.

Through the kind offices of Professor A. Rupert Hall of Imperial College, London, I was able to obtain the services of Dr. Timothy Metham, who examined a number of files in the Public Record Office at Kew.

Finally I wish to acknowledge my infinite good fortune in having a wife who adds to all the usual virtues the accomplishments of typist, critic, and source of confidence.

Chapter 1

INTRODUCTION

The first nine chapters of this book are mainly the story of the remarkable wartime achievements of a Section of a Division of the National Research Council of Canada (N.R.C.). This was the Radio Section of the Division of Physics and Electrical Engineering. On January 10, 1942 this Section became the Radio Branch, and on January 1, 1946 it was combined with the Electrical Engineering Section to form the Electrical Engineering and Radio Branch of the Division of Physics. The last chapter will deal with the work of the Branch until the end of 1946, when it became the Division of Radio and Electrical Engineering.

At the declaration of war in September 1939, the Radio Section was operating with two men employed by the Council and four others paid by the Associate Committee on Radio Research; in the latter part of 1943 it employed slightly more than two hundred people, about sixty of them professional physicists and engineers. At that time there were also about eighty men seconded from the armed services.

The Radio Section had had its origin in a decision taken on January 3, 1931 by the Associate Committee on Radio Research, one of the many unpaid committees of experts set up to help the National Research Council. This committee strongly urged that the Council "should establish a permanent branch devoted to radio research."[1] It was first necessary to find a suitable man, and Lieutenant-Colonel W. A. Steel was selected from a large number of applicants. Steel had been in charge of communications for the Canadian Corps during the First World War, and came to the Council from the Department of National Defence. He arrived on October 1, 1931, bringing with him a small amount of apparatus from the Royal Canadian Corps of Signals.

---

[1] N.R.C. Annual Report for 1930-31, p. 89.

The first investigation begun by the Section, in co-operation with the Radio Research Board of the United Kingdom, was a study of the atmospheric disturbances known as static. While this was in progress Steel resigned on January 17, 1933 and went to the new Canadian Radio Broadcasting Commission, but in the meantime, a total eclipse of the sun had occurred on August 31, 1932 and had provided the opportunity for measurements on the fading of radio signals in order to answer certain questions about the ionized layers in the upper atmosphere.

On October 1, 1933 John T. Henderson was appointed to the professional staff. Born at Montreal in 1905, Henderson graduated from McGill in 1927 and went on to get his M.Sc. in 1928. Inspired by a talk given at McGill by Professor Edward Appleton of King's College, London, he persuaded Appleton to take him as a student in 1929. In 1932 he received his Ph.D., and returned to Canada that summer to take part in the eclipse observations already mentioned, then went back to Europe for some post-doctorate study at the Sorbonne and at the Technische Hochschule at Munich. In the meantime he applied for a position at the National Research Laboratories.

The Associate Committee met soon after his appointment and decided that he should "begin work on three main problems, namely, atmospherics, measurements and standards, and wave propagation."[2] In point of fact not much time was found for work on the last of these problems, and the development of a standards laboratory was greatly delayed by the lack of funds in those years of financial depression. As to the programme of direction-finding of atmospherics, which had been expected to be very useful to the meteorologists, it became clear as the decade progressed that it would be of almost no real help to them.

But thunderstorms were not the only sources of electromagnetic energy that could be located by the use of two or more direction-finding stations. Major-General A. G. L. McNaughton had become President of the National Research Council on May 1, 1935. McNaughton, an electrical engineer by training who had developed into an exceptionally brilliant artillery officer during the war of 1914-1918,[3] saw military uses for the cathode-ray-tube direction finder. This instrument, which consists fundamentally of three special antennas, three receivers, and a cathode-ray oscilloscope, had in fact been invented in 1923, and patented by McNaughton and Steel in 1925,[4] a fact that later caused some surprise and chagrin to English scientists who invented it independently at about the same time.

---

[2] N.R.C. Annual Report for 1933-34, p. 92.
[3] See J. A. Swettenham, McNaughton, 3 vols. (Toronto: Ryerson Press, 1968).
[4] Canadian Patent No. 251024, granted June 30, 1925.

Henderson's main research in the years following 1936 was the development of the cathode-ray direction finder for aerial and marine use. Part of his time was taken up with the testing of broadcast receivers and loudspeakers for the Canadian Radio Broadcasting Commission under a contract that enabled the Council to employ three technicians for Henderson's laboratory who could not otherwise have been hired under the stringent financial controls in force at the time. As the testing only took up about a quarter of the time of the Section, this was a real gain, and before the end of 1938 Henderson was able to complete a direction-finding equipment that was taken to Camperdown, Nova Scotia. Just at this time, a new technique was being developed in England and the United States that was to put such apparatus into a secondary position. This discovery later became known by the acronym Radar, which signifies "radio detection and ranging."[5]

---

[5] For a fuller account of the pre-war Radio Section, see W. E. K. Middleton, *Physics at the National Research Council of Canada, 1929-1952* (Waterloo, Ontario: Wilfrid Laurier University Press, 1979), chap. 4.

Chapter 2

THE GENERAL PRINCIPLES OF RADAR

In 1888 Heinrich Hertz showed that the invisible electromagnetic waves radiated by suitable electrical circuits travel with the speed of light,[1] and that they are reflected in a similar way.[2] From time to time in the succeeding decades it was suggested that these properties might be used to detect obstacles to navigation, but the first successful experiments that made use of them were in an entirely different context, namely, to determine the height of the reflecting layers in the upper atmosphere.[3] One of these experiments, that of Tuve and Breit, made use of short repeated pulses of radiation, and this technique was employed in all the developments of radar that will be mentioned in this book.

Electromagnetic radiation travels in empty space at a speed of 2.998×10^8 metres per second, and in air only slightly less rapidly; we can think of its speed as very nearly 300,000 kilometres per second. This speed is denoted by the letter c. Let us suppose that a very short pulse of radiation is directed towards an object at a distance r, and that a small fraction of this is reflected back to the starting point, so that it has traversed the distance $2r$. This will take a time $t = 2r/c$. If we can measure this time we can determine an unknown distance to the target: $r = \frac{1}{2}ct$. For useful terrestrial distances t is very small; an object 15 km. away, for example, will return a signal in one ten-thousandth of a second.

In practice we need to know more about the target than its distance; we must also determine its direction. This is done by arrang-

---

[1] H. R. Hertz, *Ann. d. Phys.* 34 (1888), 551-69.

[2] Ibid., 609-23.

[3] E. Appleton and M. A. F. Barnett, *Nature* 115 (1925), 333-34. M. A. Tuve and G. Breit, *Terr. Mag. & Atm. Elect.* 30 (1925), 15-16.

ing an antenna system to project a suitable radiation pattern that can be rotated in azimuth or elevation. As may be deduced from what follows, a very great deal of ingenuity and engineering skill has been devoted to the design of radar antennas.

This is the place to mention a fundamental fact, namely, that the size of antenna needed to produce a beam of radiation of a given angular width is directly proportional to the wavelength of the radiation, or inversely proportional to its frequency. The first successful radar installations in Great Britain in the years 1935 to 1939 used wavelengths in the 6 to 15 m. band, and required very large antennas. Other equipment developed later used wavelengths of 3 m. and 1.5 m.; and in 1940 the invention of a new form of generator, the cavity magnetron, at once made it practicable to employ wavelengths of 10 cm. and even less. Nearly all the radar development at the National Research Council in 1942 and later was done at centimetre wavelengths, universally referred to as *microwaves*.

Let us now consider in a very general way the components of a radar apparatus. These comprise a powerful high-frequency oscillator, a modulator to cause it to emit pulses, and a transmitting antenna, as well as a receiving antenna and a sensitive receiver, the output of which is applied to a cathode-ray tube in the manner familiar to those acquainted with television technique. Rather early in the development of radar it was found preferable to use the same antenna for both transmitting and receiving. This is possible because the transmitted pulse of radiation is very short compared with the interval between pulses, so that the antenna can be disconnected from the transmitter during this interval, and connected to the receiving equipment. To illustrate the components of a wartime microwave radar using a common rotating antenna, Figure 2.1 is presented.[4] The TR switch disconnects the receiver during the pulses, protecting it from damage, and the ATR switch ensures that the quiescent transmitter does not short-circuit the receiver during the intervening periods. The four blocks designated "local oscillator," "mixer," "I-f amplifier," and "detector" together form the essential parts of the familiar superheterodyne receiver. The radar illustrated in this diagram would display the range and direction on a plan position indicator or PPI, in which the cathode-ray tube displays a circular section of a map with the radar station at the centre. From this centre an outward range sweep is rotated in synchronism with the rotation of the antenna. A target will be shown as a bright spot as the sweep passes it. If the cathode-ray tube is provided with a phosphor that

---

[4] From Louis N. Ridenour (ed.), *Radar System Engineering*, Massachusetts Institute of Technology Radiation Laboratory Series, vol. 1 (New York and London: McGraw-Hill, 1947), p. 7. This is probably the best general book on wartime radar techniques.

continues to emit light for more than the period of rotation of the sweep, a more or less continuous display will result, so that all targets within range will appear as on a map (Figure 2.2). Other types of display are also used.

While the principles of radar appear to be simple, the actual realization of a workable system calls for technical ability of a high order. In the first place the reflected energy received from a target such as an aircraft at a distance of some tens of kilometres may be only 10^{-18} of that radiated, so that the transmitter has to be very powerful and the receiver extremely sensitive. In the second place the entire apparatus must be rugged enough to resist weather, transport, and rough handling. The fact that very effective apparatus was produced so quickly under wartime constraints demonstrates a remarkable competence in electrical and mechanical engineering, and a no less remarkable dedication to the common cause.

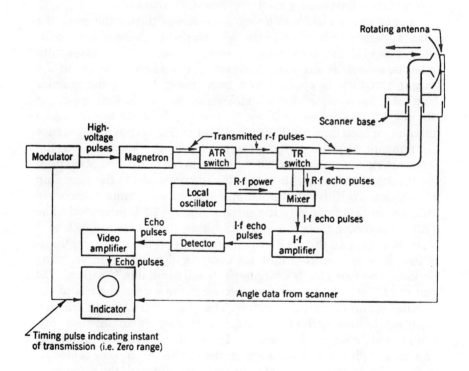

FIGURE 2.1
Components of a microwave radar set
(From L. N. Ridenour, *Radar System Engineering*, 1947.
Used with the permission of McGraw-Hill Book Company.)

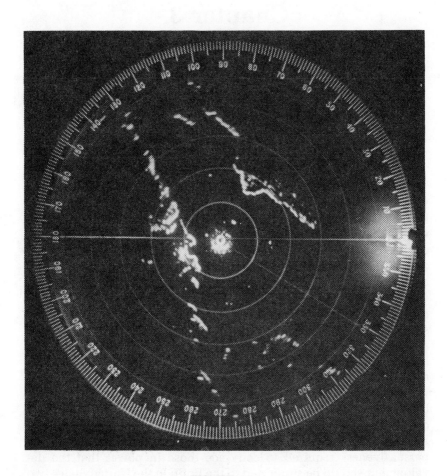

FIGURE 2.2
A PPI display

Chapter 3

PRE-WAR DEVELOPMENTS IN RADAR[1]

In the 1930s the detection of aircraft and ships by electrical methods was receiving attention in several countries, most notably the United Kingdom and the United States. The secrecy of these investigations was so complete that until the winter of 1939-40 neither of these two countries seems to have been aware that the other possessed radar equipment, although by this time they both had a good deal of it in actual operation.

The United States was the first in the field, beginning in 1930 at the Naval Research Laboratory. The British researches began rather suddenly in the spring of 1935, and were carried forward with greater and greater urgency, so that within a year a chain of warning stations had been erected on the east coast of England. This remarkable achievement was motivated by the recognition that the British Isles were extremely vulnerable to attack from the air. It was made possible by a uniquely informal organization that operated with the minimum of red tape and enlisted the abilities of some of the most brilliant physicists in the kingdom, one of whom turned out to have the persuasiveness required to convince those in power that the de-velopment should be supported to the limit. In the United States two large service organizations were involved, unused to co-operating with each other. The sense of urgency was also lacking.

As the result of questions in the British Parliament, a powerful committee was set up in January 1935 under the chairmanship of

---

[1] Much of the information in this chapter is derived from a projected publication by H. E. Guerlac that was to have been an official history of the development of radar. I have seen a microfilm of part of the typescript of this. See also R. L. Smith-Rose, *Wireless World* (March 1945), 66-70. Guerlac, a professor at Cornell for many years, is now a leading historian of science.

H. T. (later Sir Henry) Tizard.[2] This committee included Professor
P. M. S. Blackett, A. V. Hill (Secretary of the Royal Society), H. W.
Wimperis (Director of Scientific Research for the Air Ministry), and
A. D. Rowe. Wimperis immediately asked Robert Watson-Watt,
Superintendent of the Radio Department of the National Physical
Laboratory, whether radio waves could be used to destroy or cripple
attacking aircraft. Only a little arithmetic was needed to show the
impossibility of this, but the same calculations indicated that waves
reflected from an aircraft might be used to detect it. By March 1935
research was under way, and on April 10 Watson-Watt presented
several suggestions for possible lines of development. By the end of
September aircraft could be detected at ranges up to 90 km. and a
chain of warning stations was being set up around the east coast of
England. Research was quickly transferred to larger premises, and
Watson-Watt recruited about fifty physicists, mostly young and all
enthusiastic; he ran the laboratory rather like an Oxford or Cambridge
college, so that creativity was not shackled in red tape. (It is worth
noting parenthetically that the success of the National Research
Council of Canada was greatly promoted by a lack of civil-service
constraints.)

It is not my purpose to detail the very considerable advances
made by the British before 1940. It is now universally acknowledged
that the Battle of Britain could not have been won without the aid of
radar, so that in the absence of this technical development the subsequent history of the world would have been very different. Something
must be said, however, about the invention of the cavity magnetron,
the essential component in the later and more accurate equipment.

Many people with a superficial knowledge of the history of radar
have the impression that this device was a completely unprecedented
"breakthrough" that was made at the University of Birmingham soon
after the outbreak of war. It was indeed a brilliant and spectacular
invention, but work on the operation of vacuum tubes in magnetic
fields had been going on in the United States, Germany, Czechoslo-
vakia, France, and Japan for two decades, beginning with the researches of A. W. Hull of the General Electric Research Laboratories,
who took out two patents and in 1921 wrote a paper which he entitled
"The Magnetron."[3] Other devices had followed, all of which produced radiation of various wavelengths between 5 and 30 cm., but,
with one exception to be mentioned later, none of them had an output
of more than a few watts.

In 1939 a group of physicists was assembled at the University of
Birmingham under Professor Mark Oliphant with the specific task of

---

[2] See R. W. Clark, *Tizard* (Cambridge, Mass.: Massachusetts Institute of Technology Press, 1965).

[3] A. W. Hull, *J. Amer. Inst. Elect. Engrs.* 40 (1921), 715-23.

developing a high-power generator of microwaves. The group read the literature and went back to first principles. They came to recognize that the key to the problem was to combine the generator of high-frequency oscillations and the resonant circuit into a single unit. The way to accomplish this was found by two of Oliphant's team, J. T. Randall and H. A. H. Boot.

Neither of these young men had had any previous experience in the design of high-frequency oscillators, and it was probably because of this that they were little influenced by the very extensive literature on high-frequency generators, and went back to fundamentals instead. In a single afternoon in November 1939, we are told,[4] they decided on the main features of their first resonant cavity magnetron. A remarkable part of this reasoning was concerned with the size of the apparatus, which they calculated by going back to the original resonant loop of wire of Heinrich Hertz. The structure that they adopted, shown diagrammatically in Figure 3.1, is, in effect, an array of six such loops, joined by slots to a central anode cavity; but the loops are made, not of wire but out of a solid block of copper, an adequate conductor of both electricity and heat. A magnetic field is applied parallel to the axes of the resonators, and at a critical field-strength the electrons spiral out from the cathode, passing the slots at grazing incidence and covering the distance between two successive slots in half the period of oscillation.

After much delay in getting the components made, the cavity magnetron was operated for the first time on February 21, 1940 from a direct-current power supply. They found that the tube was radiating about 400 watts at a wavelength of 9.8 cm., and that the frequency was remarkably stable. By June 1940 sealed-off tubes were being produced commercially in England, and within a few months, magnetrons operating on pulsed voltages were giving peak powers of 10 to 50 kW. at efficiencies of 10 to 20 per cent. One of these magnetrons was brought to North America by the famous British Technical Mission (the Tizard Mission) in September 1940, the visit of which will be discussed in Chapter 5.

There is no doubt whatever that Randall and Boot invented the cavity magnetron. Nevertheless they were not the first to do so. In 1936 and 1937 N. F. Alexseev and D. D. Malairov in Leningrad built various cavity magnetrons, one of which, with four loops, gave 300 watts of continuous radiation at a wavelength of 9.0 cm. with an efficiency of 20 per cent. This was published in the U.S.S.R. before the war and in translation in the U.S.A. in 1944.[5] Apart from the number

---

[4] By Guerlac in the manuscript referred to (see footnote 1 of this chapter). See also J. T. Randall and H. A. H. Boot, *J. Inst. Elec. Engrs.* 93 (1946), part IIIa, 182-83.

[5] Alekseev and Malairov, *Proc. Inst. Radio Engrs.* 32 (1944), 136-39. I am indebted to A. E. Covington for this reference.

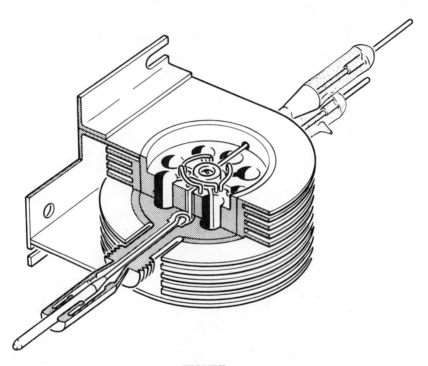

FIGURE 3.1
Diagram of the cavity magnetron

of loops, the structure of the anode was similar to that of the British magnetron. But no application of this promising invention is mentioned, and the very fact of its publication is puzzling.

It is interesting that the general use of microwaves about 10 cm. long in much wartime apparatus apparently stems from the arbitrary decision made in the laboratory at Birmingham to try to make a magnetron to generate that wavelength. The region of the spectrum near 10 cm. became known as the S-band. Later, as we shall see, even shorter waves were employed.

Chapter 4

THE RADIO SECTION IS INTRODUCED TO RADAR

As early as the autumn of 1938, Major-General A. G. L. McNaughton, President of the National Research Council, who saw clearly that war was imminent, began to discuss with the Department of National Defence the detection of aircraft by electrical means. In the spring of 1939 the Air Ministry of Great Britain thought it advisable to make known to the Dominions, under strict security, the considerable progress that the British had made in this field. On March 10, 1939 Air Vice-Marshall Croil, Chief of the Air Staff in Canada, asked President McNaughton to send a physicist to Great Britain. On March 18 J. T. Henderson sailed from Halifax.

When he arrived in London he was met by Squadron Leader F. V. Heakes, the Royal Canadian Air Force (R.C.A.F.) Liaison Officer in the United Kingdom. In the company of representatives from Australia, New Zealand, and the Union of South Africa, they visited research laboratories, radar stations, and factories, and in April sent a preliminary report[1] back to Canada that shows by the amount of information concisely presented that Henderson at least was an acute and knowledgeable observer.

The report begins with a "general description of R.D.F."[2] I cannot do better than quote its first paragraphs in order to provide a ready reference to most of the types of equipment mentioned later in this book:[3]

---

[1] PRA-6, F. V. Heakes and J. T. Henderson, "Electrical Methods of Fire Control in Great Britain."

[2] I.e., radio direction-finding, the term used then and for some time afterwards by the British for what we now call radar.

[3] Not all the abbreviations used in the United Kingdom for these types were adopted in North America. See Appendix A.

1. Definition

R.D.F. is a system of radio detection and position determination of aircraft and ships from a single station based on the principle of reflected radio impulses using ultra short wave-length. The reflected echoes are registered on cathode-ray tubes and the distance, azimuth and elevation of the aircraft or ship are determined.

2. Functions

R.D.F. is designed to fulfill the following functions with respect to the defence of Great Britain:-

(a) Detection and determination of altitude and direction of approach of raiding aircraft or formations, with a view to interception of the raiders, warning to the ground defences and air raid warning to the civil population. For this purpose, a chain of stations has been erected along the South and East coasts of Britain and other strategic areas, giving complete coverage of all approaches over the sea. This type is known as *C.H.*

(b) Detection and determination of ships' position from reconnaissance aircraft under all conditions of visibility, day or night, the device increasing the scope of reconnaissance aircraft, giving a lateral coverage, each side, of thirty miles. This type is known as *A.S.V.*

(c) Detection and determination of direction of approach and range of aircraft by A.A. Artillery, and directing Searchlights, the sets being in close proximity to the guns and the information obtained being fed directly into the predictor. This type is known as *G.L.*

(d) Detection and determination of altitude, direction of movement and position of raiders by use of small mobile units which can be used as intermediate to type *C.H.* (described in (a) above); also for inland R.D.F. and for use with Expeditionary Force. This type is known as *M.B.*

(e) Detection of ships and determination of position from shore stations, having a somewhat restricted range depending on height of station above sea level. This will be used by coast defence artillery. This type is known as *C.D.*

(f) "Homing" of fighter aircraft, particularly on aircraft which might be used for jamming the R.D.F. system. This type is known as *A.I.*

(g) Detection and determination of position of ships from ships. This type is known as *S.S.*

(h) Detection of aircraft from ships, another naval development on which we have not yet been able to get details. This type is known as *S.A.*

They go on to remark that very different equipment is needed to fulfill these various requirements. After a few paragraphs on the way in which the CH stations are used, they deliver the following opinion, which emphasizes the overwhelming impression made by this new development even in its early stages: "R.D.F. represents such an advance in the science of war that it is difficult, at this point, to determine or approximate its strategical or tactical effects in the future."

Part II of the report consists of brief descriptions of the various types of equipment shown to Heakes and Henderson, with emphasis on the type known as CH (chain home), the early-warning radar. A chain of such installations had already been set up along the east coast of England, working on wavelengths of 5 to 13 metres and hence requiring very massive towers as much as 350 feet high to support the antennas. The second most important type was ASV (air-to-surface vessel), a relatively light-weight equipment installed in aircraft, mainly to enable reconnaissance aircraft to detect ships at sea. This used a wavelength of one metre. A third type was GL (gun-laying) radar to assist in the aiming of anti-aircraft guns. It eventually turned out that a very large part of the Canadian war effort in the field of radar was in the improvement and production of this type of equipment.

Part III deals with the strategic and tactical implications of radar, first to Great Britain, and then to Canada. Part IV is entitled "Particular Application to Canadian Defence," and presents two proposals for the deployment of radar equipment on the east and west coasts of Canada. The writer of this section, presumably Heakes, realizes that even the lesser of the two schemes might be thought unreasonably expensive. In a hurriedly written Part V the possible cost was estimated, on the assumption that the equipment could be bought from Britain, an assumption recognized as doubtful.

On his return to Canada in June 1939 Henderson wrote a second report [4] presenting a good deal of technical information that he had picked up. Nobody knew that within a few months much of this would be made irrelevant by the invention of the magnetron, which was to make it possible to reduce by an order of magnitude the size and weight of most types of radar, at the same time greatly improving the performance of these devices.

In June the Canadian Chiefs of Staff, impressed by the possibilities of radar, submitted a plan to expend four hundred thousand dollars for the purchase of British radar equipment to defend the coasts of Canada. It scarcely matters that this plan was not accepted by the Government, for so much apparatus could certainly not have been obtained. What was more important is that requests for additional funds for research fell on deaf ears, even after the outbreak of war.

In July 1939, after consulting the Chiefs of Staff, McNaughton had set out a list of the ways in which the National Research Council could help the armed forces in the matter of radar:[5]

---

[4] PRA-7, J. T. Henderson, "Electrical Methods of Fire Control in Great Britain (Report No. 2)."

[5] I have used the term radar, now more familiar than the abbreviation RDF used in the documents of the period.

(1) to act as general consultants to the Department of National Defence;

(2) to adapt British designs to suit Canadian practice;

(3) to do research on radar for the Department;

(4) to assist with the installation and calibration of new types of equipment; and

(5) to train a nucleus of technical ratings and expert operators for the Department of National Defence.

At this time it was not expected that the radio laboratory would be engaged in the development of new types of equipment, and it is clear that McNaughton saw its role as serving the armed forces in the most practical and immediate way possible. In July the sum of $105,000 was requested, to pay for additional equipment and to hire twenty-three additional staff members, but the expenditure was not approved. In fact all that could be obtained during the fiscal year that ended on March 31, 1940 were some small internal reallocations of funds amounting to less than $6,000, the transfer of some equipment from the Department of National Defence, and the seconding of four men from the Army to work in the Radio Section. Meanwhile in August 1939 General McNaughton was in England and proposed that radar equipment should be manufactured in Canada. Partly because of concern about secrecy, but also because the British characteristically underrated Canadian abilities,[6] the proposal was rejected.

Dr. R. W. Boyle, the Director of the Division of Physics, had been in Germany when war broke out, and had escaped to England with some difficulty. Dr. Boyle, who, by a remarkable coincidence, had made a similar escape at the beginning of the 1914-1918 war, and had then done valuable war research in England, went to see his friends in the British Services and obtained additional details about radar and other secret matters.[7] When he returned to Canada in October he tried to get funds to enlarge the laboratory, but none were immediately forthcoming.

On October 17, 1939 McNaughton took leave from the Council to take command of the First Canadian Division, and C. J. Mackenzie, a distinguished engineer, was appointed acting President. C. J. Mackenzie, who is ninety years old as I write, is a very great Canadian. A graduate in civil engineering from Dalhousie and Harvard, he served with distinction in the 1914-1918 war, and in 1921 became Dean of Engineering at the University of Saskatchewan. He had been a member of the Honorary Advisory Council[8] since 1935 and had ac-

---

[6] This is clear from reports by A. V. Hill, Secretary of the Royal Society, at the Public Record Office, London, AVIA 22/2286, June 17 and 18, 1940.

[7] His report was duplicated as PRA-5, "Report on Trip to England."

[8] This body is usually known as the National Research Council. It reported to the Sub-committee on Scientific and Industrial Research, a committee of the Privy Council.

quired a good deal of knowledge of the operation of the Laboratories because of his work on the Review Committee. When McNaughton decided not to return to the presidency of the Council in 1944, Mackenzie assumed the office.

For the new fiscal year of 1940-41 the Treasury passed a special war appropriation, "P.I.-100, Radio (secret)—$59,800.00."[9] This increased budget permitted a rapid increase in activity, in fact about 50 per cent in the next three months, if we judge by the growth in manpower employed in the laboratory. Much more financial help was to come. In June, shortly after the evacuation of Dunkirk, a group of Canadian businessmen, deeply shocked by the events in Europe and scarcely less so by the inaction of the Government, made several hundred thousand dollars available to help the prosecution of the war, and with no other restrictions. The way in which this money was allocated was described by C. J. Mackenzie in a letter written to General McNaughton on July 25, 1940:

> About three weeks ago Mr. Duncan, acting deputy minister of National Defence for Air, phoned me that he was calling a meeting to discuss how certain monies which had been contributed to the government might best be spent. At the meeting were present Mr. R. A. C. Henry representing Mr. [C. D.] Howe, Mr. [H. L.] Keenleyside representing Dr. [O. D.] Skelton, Colonel [Alan A.] Magee representing Colonel [J. G.] Ralston and one or two others. Mr. Duncan explained that John Eaton and Sir Edward Beatty had each offered about a quarter of a million dollars as a contribution to Canada's war effort. Mr. Duncan thought that instead of putting the money into the general revenue it would be better if it could be allocated to specific projects and the various people present at the meeting were asked to suggest suitable projects. After several suggestions were made, I was called upon and presented briefly the way in which the Council had been working, showing them that in all the projects such as equipping the gauge laboratory, preparing for the manufacture of optical glass, erection of the Aeronautical Building, work on radio, chemical engineering, etc., we had anticipated the nation's needs by several months. I pointed out that had we not so functioned, it would be almost impossible at the present time to equip laboratories such as the gauge department and that conditions would be very serious. In about ten minutes all other projects were discarded and it was decided to suggest to the donors that they make their contributions to the National Research Council for war technical developments.[10]

There was no longer a shortage of money—only of manpower and materials. This splendid gift, affectionately named the "Santa Claus Fund," finally amounted to more than a million dollars. It enabled the

---

[9] Honorary Advisory Council, Meeting 130, June 3, 1940, minute 35, Exhibit N.
[10] Mel Thistle (ed.), *The Mackenzie-McNaughton Wartime Letters* (Toronto: University of Toronto Press, 1975), pp. 42-43.

population of the Radio Section to rise from 21 in April 1940 to 105 a year later, not counting people seconded to it from outside.[11]

John T. Henderson had not waited until those in power could be convinced of the value of radar. Tremendously impressed by what he had seen and heard in England, and with the encouragement of the Canadian Army, he began work on an equipment for coast defence in September 1939. He made use of available components and chose a wavelength of 60 cm.[12] This was a Canadian effort that seemed particularly appropriate to the Chiefs of Staff, especially as their request for funds to purchase such equipment from Great Britain had been denied. Henderson was also asked by the R.C.A.F. to work on a radar that could be mounted in an aircraft to detect surface vessels.[13]

When war was declared the professional staff of the Section consisted of Henderson, J. W. Bell, D. W. R. McKinley, and H. Ross Smyth. F. H. Sanders, who had been administrative assistant to the Director of the Division of Physics and also had much experience in ultrasonics, was transferred to the Section in February 1940, and G. R. Mounce joined at the same time. With a rapidly increasing number of technical and workshop assistants, these men worked hard through the summer of 1940. In early autumn the arrival of a British Technical Mission (nominally in the United States, but also in Canada) changed the state of affairs greatly.

---

[11] John T. Henderson, "Progress Report for Period June, 1939 to January 1, 1942" (mimeographed), Appendix B.

[12] Further developments in coast-defence radar will be related in Chapter 8.

[13] See Chapter 9.

Chapter 5

LIAISON AND THE TIZARD MISSION

During the early months of the war there was little liaison between the National Research Council and British scientists and engineers. Indeed, few people in the United Kingdom were aware that the Council existed. One who had learned something about Canadian potential for research and development was the physiologist A. V. Hill, one of the secretaries of the Royal Society from 1935 to 1945. R. W. Boyle had met Hill during his stop in England in September 1939, and Hill had been impressed by what Boyle told him.[1] It is not clear to what extent Boyle discussed his visit with C. J. Mackenzie and J. T. Henderson, but there is a cryptic entry in Mackenzie's diary for December 8, 1939 that states that he had seen Boyle and Henderson about RDF.

In the spring of 1940 A. V. Hill was sent to Washington as supernumerary air attaché in the hope of getting American scientists into the war even though their country was still neutral. Remembering Boyle's visit, he came to Ottawa and on May 1 he was in Mackenzie's office.[2] This was the beginning of real liaison with the United Kingdom. He talked to the scientific staff and to colleagues in Toronto and Montreal. He consulted Sir Gerald Campbell, the British High Commissioner. After he returned to London he wrote a memorandum entitled "R.D.F. in Canada and the United States."[3] This remarkable and forthright document is dated June 18, 1940, and its first five paragraphs deal with Canada, beginning with the "small but active group" under J. T. Henderson, who told him that he had the names of

---

[1] Interview of A. V. Hill by D. J. C. Phillipson, September 1, 1976.
[2] C. J. Mackenzie's "War Diary." I am indebted to Dean Mackenzie for permission to consult this still unpublished document.
[3] London, Public Records Office, file AVIA 22, 2286.

300 men who "would be capable of undertaking R.D.F. work at once."
The universities were developing training schemes. The next paragraph reads as follows:

> At present these schemes are going rather slowly owing to the absence of
> official support from the Canadian Government and the lack of money
> and equipment. *If the Canadian Government were asked by the British
> Government to develop the work on a much larger scale it would almost
> certainly agree to do so, and all these plans would come together into an
> organized whole.*

The italics are Hill's. He goes on to note the friendly relations between
scientists in Canada and the United States. As to the United States, he
goes into detail about what he has been able to see behind the curtain
of secrecy, and ends with an earnest plea and a detailed plan for the
exchange of information between the United States and the United
Kingdom. Hill says:

> If the proposal is accepted, the next steps will be
>   (a) to send a mission to the U.S.A.
>   (b) to inform the Canadian Government what has been done, and
>   (c) to ask that representatives of the Canadian National Research
> Council and of the Canadian Services should collaborate with the Mis-
> sion on the developments to be expected.

In his final paragraph he says that it is obvious "that Canada may have
to exercise a fundamental role in Imperial Defence, and that we
should do everything possible to get things going there and to use and
improve their resources." This was written a fortnight after the
evacuation from Dunkirk.

As we know, his plan for a technical mission was accepted and
bore fruit in less than three months. What is not usually pointed out in
the accounts of those times is the importance that Hill attached to the
Canadian potential.

On the previous day he had signed another memorandum[4] entitled "Research and Development for War Purposes in Canada." This
was equally forthright: "Everything I saw and heard [in Canada]
convinced me that there had been a grave lack of imagination and
foresight on our part in failing to make full use of the excellent
facilities and personnel available in Canada." He pointed out that
research and development in Canada would be safe from bombing by
the enemy, that there would be plenty of room for trials by the
Services, and that Canadian scientists and engineers would have
access to technical help from the United States.

The part that Hill played in these events is little known,[5] and
should be remembered. In the same memorandum he urged that a

---

[4] P.R.O., ibid.

[5] There is a short obituary by D. J. C. Phillipson in *Science Forum* (December
1977), p. 27, which emphasizes this.

scientific liaison officer of high attainments should be sent to Canada at once. The physicist R. H. Fowler in fact arrived in Ottawa on September 8, 1940.

The idea of an exchange of information with the United States, strongly supported by Sir Henry Tizard, was the subject of intense discussion in the United kingdom in the spring and summer of 1940.[6] The suggestion aroused much opposition, largely on the grounds of security, but also because of the British belief, held even by some important scientific men, that the Americans could teach them nothing. After political and bureaucratic manoeuvres that are now hard to credit, the final approval for the mission was given by Churchill only on August 9. Tizard and Group Captain Pearce left England by flying boat on the fourteenth. The remainder of the members of the Mission sailed from Liverpool in the *Duchess of Richmond*, arriving at Halifax during the first week of September and reaching Washington on the seventh.

Tizard had insisted that the Service members of the Mission should be "Serving Officers who had recently been in action and who had had some operational experience of radar or similar aids."[7] The Mission was therefore composed of: Sir Henry Tizard; Captain H. W. Faulkner, from the Navy; Colonel F. C. Wallace, from the Army; Group Captain J. L. Pearce, from the Air Force; Professor J. D. Cockcroft of Cambridge; Dr. E. G. Bowen; and Mr. A. E. Woodward Nutt, Secretary. Cockcroft and Bowen were the technical experts on radar. Of the others, Colonel Wallace is of particular interest to us. He had been in charge of anti-aircraft radar with the British Expeditionary Force, and had had the heartbreaking task of blowing up all the radar equipment on the beach at Dunkirk. As we shall see, he was in a large measure responsible for the success of the Canadian programme of research and production in the field of radar.

Tizard, probably influenced by Hill's reports and conversations, came to Ottawa for several days before he went to Washington. He came into C. J. Mackenzie's office on August 16, and the two men "immediately became intimate friends, as we seemed to talk the same language and reacted to most situations alike."[8] Mackenzie introduced him to everyone available who could be useful, in the Services, in politics, and in the universities. It appears that Sir Henry was so greatly impressed by the good he might do in Canada that after a brief visit to Washington he came back to Ottawa on August 26.[9]

The rest of the Mission landed at Halifax on September 6 and went directly to Washington, where the official meetings with the

[6] R. W. Clark, *Tizard*, chap. 11.
[7] Ibid., p. 259.
[8] C. J. Mackenzie to Sir Harold Hartley, quoted by Clark, *Tizard*, pp. 261-62.
[9] Mackenzie, "War Diary." I have not seen this fact reported elsewhere.

Americans began on the tenth. With Colonel Harry Lettson, the Canadian military attaché, Mackenzie arrived in Washington on the twelfth, and the next day telegraphed S. P. Eagleson, the Secretary of the Council, "Have Henderson come Washington arriving Sunday."[10] Sunday was the fifteenth, and Henderson went to Washington and to the meetings. Air Vice-Marshal E. W. Stedman represented the R.C.A.F., and the Deputy Chief of General Staff, Brigadier Kenneth Stuart, was the delegate from the Canadian Army. In the following week J. W. Bell of the Radio Section went to Washington for a few days.

The Mission, or parts of it, travelled widely, going to laboratories in New Jersey, Dayton, Ohio, and Indianapolis. But as far as radar is concerned the really important event was the exhibition of the cavity magnetron on September 30. This at once opened up new vistas to the American engineers, who had been getting on with ten watts of power and were suddenly presented with ten kilowatts. As a direct result of this disclosure and discussions with Cockcroft and Bowen the famous microwave laboratory at the Massachusetts Institute of Technology, later known as the Radiation Laboratory, was founded.

On the first of October, in Washington, Tizard dictated a secret memorandum entitled "Future Liaison with the National Research Council of Canada."[11] Some paragraphs of this deal with radio and radar:

> 1. When I return to England, I shall urge that first-class scientific personnel should be sent to the National Research Council to assist in the rapid development of radio research, including radio navigation.
>
> . . . .
>
> 4. In general, after conversations with Dean C. J. Mackenzie . . . and others, I think it desirable to move as much research as possible to Canada, particularly when the nature of the research is such that it may be followed up by production of such equipment in Canada or the U.S.
>
> . . . .
>
> 10. Professor Cockcroft and Dr. E. G. Bowen should spend some time in Canada in consultation with the National Research Council before they return to England in order particularly to assist in developing the program of radio research and similar work, particularly work which involves full-scale flying.

It is evident that Sir Henry believed that the Canadians needed help with radio research, and was proposing to move a good deal of the British research to Canada. It is probably as well that this was not done, in view of the remarkable developments that took place in the British Isles.[12]

---

[10] N.R.C. File B3.25.1.51.

[11] Ibid.

[12] See the proceedings of the Radiolocation Convention held in London on

Tizard returned to England on October 5, 1940. The rest of the Mission, with the exception of Wallace and Bowen, returned later that year. On October 12 Cockcroft wrote to Mackenzie to say that Bowen, Wallace, and he hoped to arrive in Ottawa on the twentieth.[13] They did, and on the twenty-second and twenty-third there were meetings at which they briefed members of the Canadian Armed Services. Cockcroft sailed for England from New York some time later in November; Bowen continued to engage in liaison work between the United Kingdom and the United States; and Wallace remained in Canada. He was apparently in a peculiarly indefinite position in the British Army after Dunkirk; in any event he was loaned to the Canadian Army and by them to the National Research Council, with results that will become clear in subsequent chapters.

Professor R. H. Fowler, as the official scientific Liaison Officer from the United Kingdom, kept himself informed about the work that was going on, both in Washington and in Ottawa. On November 4, 1940 he wrote at some length to Mackenzie about various aspects of the radar development, emphasizing co-operation between the Canadian and United States laboratories.[14] This co-operation will be referred to frequently in this book as we deal with the various kinds of apparatus.

Before Cockcroft left Washington he wrote the following memorandum:

Acting on the proposal of Dean Mackenzie, we put forward the following suggestions for the future radio program of N.R.C.
1. Assistance to Research Enterprises[15] in establishing the manufacture of ASV equipment in Canada.
   The Lockheed Hudson machine now attached to the experimental flight should at once be fitted with the ASV aerials sent from England so that the system can be demonstrated in Canada at the earliest possible moment. We consider that this equipment would be of the greatest assistance to the Halifax Squadron.
2. Development of AI set on 10 cms. We understand that it is your intention to detach a group to work in the closest cooperation with the Microwave laboratory of N.D.R.C. to be established at M.I.T.
   It is important that a suitable machine to carry the equipment should be made available with the R.C.A.F. experimental flight.
3. Construction of Prototype GL-3 set working on 10 cm. wavelength.
   A draft specification of this set is appended. We have been promised the latest information from England and hope that McKin-

March 26 to 29, 1946, in *J. Inst. Elec. Engrs.* 93 (1946), Part IIIA. In this 778-page document Canada is represented by two papers by Professor W. H. Watson of McGill University. The British achievements in radar are set out in much detail.
[13] N.R.C. file B3.25.1.51.
[14] Ibid.
[15] For Research Enterprises, Ltd., see p. 41 below.

ley[16] will also bring back fullest details. We suggest that the preliminary steps recommended on page 5 of the specification should at once be taken.

This development may also form part of the Microwave program of N.D.R.C. but is considered that Canada might take the initiative in this matter.

4. Development of a long-range radio navigation system.

It is the intention of N.D.R.C. to develop the grid-laying system which should enable a pilot to locate himself with an accuracy of better than one mile with ranges of up to 500 miles.

The work will involve the construction of high power pulse transmitters, using the British Nav 1 megawatt valves and the construction of transmitting towers; also, the construction of suitable receivers and cathode ray tube display systems.

Dr. Loomis intends to erect three stations in the first instance, one of which might be in the region of Nova Scotia. It is suggested that this might also be a joint project and that possibly Canada might erect one of the stations. Dr. Bowen will prepare a detailed memorandum on the subject.

5. Development of a Blind Landing system suitable for war purposes.

It has been proposed that a useful blind landing system could be prepared with very little equipment by placing IFF beacons on air fields to operate in conjunction with the 10 cm. AI when developed. N.D.R.C. proposed to follow this up.

It has also been suggested that blind landing could be carried out by using a ground RDF set to direct the pilots in approaching the aerodrome.

6. Production of 50 cm. CD sets for the control of coast defence batteries in Canada and the British Dominions generally.

We consider that the present work should be continued but that a 50 kW. transmitter using the British Micropup valves should be built.

Care should be taken that the range data is provided in a form suitable for the battery, and it is important that McKinley should bring back the latest British proposals on this subject, together with full details of their Prototype receiver.

7. 150 cm. CD Sets.

We suggest that the 150 cm. CD set and the British CHL set should be erected on the Coast to act as medium range (70 miles) warning sets and to provide operational training.

8. Long-range Warning Sets.

We understand that it is in the intention of N.D.R.C. to build a three-meter long-range warning set incorporating our megawatt valves and a radial time base display system. They would use the high towers of the long-range navigation system to mount the aerials. We suggest that this might be a joint project.

---

[16] See p. 25 below.

9. Searchlight Director.

        We consider that N.R.C. should develop a searchlight director for Canadian use since this would make unnecessary the purchase of sound locators.

        Such a director could either follow the British design on 1.5 meters or be developed for 10 cms. We think that the latter is the best policy since one could then use a common identification system for GL, SLC, and AI.

With the exception of its fourth and fifth paragraphs this document might almost serve as a summary of much of the work of the Radio Section during the war, although its different items were pursued with various degrees of priority.

In May 1941 Fowler was replaced as United Kingdom Liaison Officer by Sir Laurence Bragg, then Director of the Cavendish Laboratory, and when Bragg returned to England in the autumn he was succeeded by Professor G. P. Thompson. For liaison with scientific bodies in the United States the Council obtained the services of Professor A. G. Shenstone, a Canadian citizen who left his Princeton professorship at considerable personal sacrifice to serve his country in this way.

At the request of the R.C.A.F., D. W. R. McKinley was sent to England for the last months of 1940, returning in February 1941. He was able to send back much valuable current information as well as some samples of equipment. During this period and later the acting President was being urged by various people in both England and Canada to establish a liaison office in London. At first Mackenzie did not believe that anyone of sufficient seniority could be spared,[17] but in August 1941 Dr. L. E. Howlett, head of the Optics Section of the Division of Physics, was sent to London to initiate the liaison office which for a quarter of a century, under several liaison officers, rendered a multitude of useful services.

Before leaving the subject of liaison, it cannot be emphasized too much that during the entire period covered by this book the Americans co-operated in every possible way with the engineers and physicists of the National Research Council, especially by encouraging visits by members of the staff to the great Radiation Laboratory at the Massachusetts Institute of Technology, where most of the radar research in the United States was concentrated. There is no doubt whatever that this liaison made a very great contribution to the success of the programme both in Canada and in the United States.

---

[17] There are many documents in N.R.C. file B3.25.1.53.

Chapter 6

THE WARTIME ORGANIZATION

6.1 Radio Section and Radio Branch

We have seen, at least in outline, what was expected of the National Research Council of Canada in the field of radio and especially radar. In later chapters we shall examine the accomplishments. Now we must address ourselves to the administrative and practical arrangements that made these achievements possible. In an organization that grew to a strength of more than two hundred people there were naturally some personal conflicts, although as far as I can tell at this distance in time, these were remarkably few in the organization itself. But matters were complicated by the requirement that many of the ideas developed in Ottawa had to be put into production in another organization that eventually grew to be much the larger of the two, Research Enterprises, Ltd. (R.E.L.). We must see how R.E.L. was formed and deal with its interaction with the National Research Council in the field of radar. Closer to home there were excellent relations between the workers in radar and the Armed Services, probably in great part a result of the talents and enthusiasm of General A. G. L. McNaughton in the years before the war.

In previous chapters we have referred to the Radio Section. From now on the term Radio Branch will also appear. The organization, at first only a Section of the Division of Physics, came to include a number of groups working on various aspects of the various problems. Beginning on August 16, 1941, a weekly meeting of the heads of these groups was held each Saturday morning. On January 10, 1942, after a reorganization that will be discussed later, this assembly was named the Co-ordinating and Management Committee, the groups became Sections, and what had been the Radio Section became the Radio Branch. The C. and M. Committee met weekly, usually on Saturdays, for the entire period dealt with in this book. The minutes have been preserved. At first they contain records of a good deal of technical discussion, but by the middle of 1942 they are almost entirely devoted to administrative concerns of the most diverse kinds, largely housekeeping matters and problems related to personnel. With one exception the members of the Committee were all without experience in administration.

The Branch was originally divided into seven sections, as follows: Navy Section, headed by H. R. Smyth; Army Section, W. Happe, Jr.; Air Force Section, D. W. R. McKinley; Special Research and Development Section, J. W. Bell; Microwave Section, F. H. Sanders; Mechanical Engineering Section, H. E. Parsons; Workshop Section, D. L. West. Later, with the astonishing proliferation of technical documents from the allied countries—15,000 had been received by August 1945—a Technical Information Section under W. J. Henderson was added in August 1942. The Special Research and Development Section comprised an Antenna Group and a Circuit Group.

The Branch was, of course, not autonomous. It was part of the Division of Physics and Electrical Engineering, if only nominally. In fact it was directed by a Radio Board, appointed by the Honorary Advisory Council at its 138th meeting, on March 19, 1942. I quote from Exhibit K attached to minute 16 of that meeting:

RADIO BOARD

Acting President, National Research Council (C. J. Mackenzie)	— who will generally exercise financial control for the Board.
Director, Division of Physics and Electrical Engineering (Dr. R. W. Boyle)	— who will generally exercise scientific and technical control for the Board.
Managing Director of the Board (Col. F. C. Wallace)	— who will be the executive officer of the Board and be responsible for all detailed administrative activities for the Board.

The document goes on to list the Sections and their leaders as detailed above. This organization had in fact been decided on in January 1942.[1] The inclusion of R. W. Boyle was purely because of his position as Director of the Division. By this time C. J. Mackenzie had arrived at the opinion that Boyle was entirely ineffective.[2] It is clear that all

---

[1] See p. 32 below.

[2] This is evident from Mackenzie's diary. See also W. E. K. Middleton, *Physics at the National Research Council of Canada, 1929-1952* (Waterloo, Ontario: Wilfrid Laurier University Press, 1979), chap. 5.

important internal decisions were made by Mackenzie and Wallace. The relations of the Branch with the Armed Services had already had the attention of the Honorary Advisory Council, who on July 16, 1940, had appointed the following "Secret Panel" of the Associate Committee on Radio Research: Commander J. L. Houghton, representing the R.C.N.; Colonel H. E. Taber, representing the Canadian Army; Wing Commander R. McBurney, representing the R.C.A.F.; the acting President of the N.R.C.; the Director of the Division; Dr. J. T. Henderson, Secretary.[3] A little later this became known as the "RDF Committee."

6.2 Population Explosion

The spectacular increase in the activity of the Radio Branch cannot be better displayed than by a diagram (Figure 6.1) taken from the "War History of the Radio Branch."[4] Starting from a half-dozen employees at the outbreak of war, the staff rose to slightly over 200 by the end of 1942, and remained remarkably constant thereafter until the defeat of the Axis powers. Of these, about sixty were in the professional grades. The number of laboratory technicians reached a maximum early in 1942, and then declined because it was seen that there was a greater demand for their services in the workshop, to which some of them were then transferred. In the next three chapters we shall see that circumstances often forced the Branch to build complete radar sets for the Armed Services so that they might be available before R.E.L. could get into production. As a result of these demands the workshops of the Section had to grow until, in the last years of the war, about fifty toolmakers, machinists, and fitters in two shifts were employed in the machine shop and sheet-metal shop. The difficulty that was caused by the distance between the main building in Sussex Street and the field station was serious. The best that could be done was to keep the main shop at the Sussex Street building and to establish smaller shops at the field station. At the main building there was also a paint shop and a plating shop. For some reason, all these shops as an organization were known as the model shop. There was also a carpentry shop at the field station and a "large vehicle assembly shop" at the John Street Annex to the main building. This last was made necessary by the development of the GL IIIC radar. All these workshops were managed by D. L. West until his death at the end of 1944, when he was succeeded by I. L. Newton.

The Services frequently lent various tradesmen to help in rush jobs, and it will be seen from the figure that through most of 1944

---

[3] Honorary Advisory Council, 131st meeting, minute 11.

[4] "War History of the Radio Branch," N.R.C. Report ERA-141 (Ottawa, August 1948), mimeographed, many illustrations.

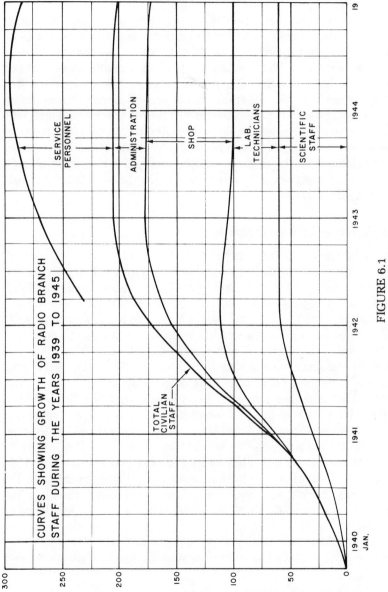

FIGURE 6.1
Growth of the Radio Branch

more than eighty servicemen were employed in this way, not to mention the highly qualified engineer officers from the Services who constantly consulted with the scientific staff of the Branch in order to make clear the operational requirements.

There were no really serious labour problems, except for understandable discontent among the technicians and machinists in 1941, when they found themselves being paid much less than they could get elsewhere. C. J. Mackenzie used his considerable powers of persuasion to convince the Minister of Finance, the Honourable J. L. Ilsley, to remedy this situation,[5] but it proved impossible to get increased salaries for J. T. Henderson and the other scientific men who had been in the programme since the outbreak of war. Henderson, as Mackenzie noted in a letter to McNaughton,[6] was responsible for the very existence of the Branch, which then had a budget of about a million dollars a year. It was undoubtedly inequitable that he should be paid only $3,300.

6.3 A Necessary Reorganization

But although Mackenzie had a high and well-justified opinion of Henderson's scientific ability, energy, and hard work, it became evident to him even before the end of 1940 that administrative difficulties were developing in the radio laboratory. The first of these was due to a natural reluctance on the part of R. W. Boyle (although he really did not care for administration) to see himself entirely excluded[7] from the control of this vital part of the Division of Physics. There is a brief entry in Mackenzie's diary on November 25, 1940: "Dr. Boyle telephoned re difficulties in dual control in radio lab. Will see Fowler about this." This is strange for two reasons: first that Boyle would discuss such a matter in a phone call rather than a visit, and secondly, that the United Kingdom Scientific Liaison Officer should be the man to whom the acting President would turn for advice. What advice he got I do not know, but what happened was that in February 1941 Colonel F. C. Wallace,[8] who had been the military member of the Tizard mission and had remained in Canada on loan to the Canadian Army, was attached to the National Research Council at Mackenzie's request.[9]

Ostensibly Wallace was to engage in liaison with the U.S. Armed Services, but it turned out that this was the least of his immensely

[5] Mackenzie, "War Diary," December 4, 1941.

[6] On August 2, 1941. See Mel Thistle (ed.), *The Mackenzie-McNaughton Wartime Letters* (Toronto: University of Toronto Press, 1975), pp. 85-86.

[7] Interviewed by D. J. C. Phillipson on September 13, 1975, F. C. Wallace said that the laboratory was completely autonomous when he arrived early in 1941.

[8] He was promoted to Brigadier in October 1943.

[9] Honorary Advisory Council, 134th meeting, minute 25.

useful activities. He must have been involved in the radio laboratory almost at once, because on April 3, 1941 Mackenzie noted that Wallace was "getting a lot of valuable work done" there. By this time it was quite clear to Mackenzie that Henderson's sterling qualities simply did not include the ability to manage a large staff. Wallace, who had managed a textile business before the war and was an extremely able administrator, apparently took gradual control of the Branch, with Mackenzie's blessing, during the summer of 1941. How this was done I have not determined, but on August 15 the acting President called Henderson and Wallace into his office and, in the words of his diary, "suggested that Col. Wallace take over the management of the administrative aspects of radio. Henderson is enthusiastic. It may work." Three days later Mackenzie simply told Boyle what he had done, and Boyle "thoroughly agreed";[10] probably he was relieved.

It worked, as far as administration was concerned. Weekly meetings of the heads of groups, with Wallace in the chair, began on August 16—the morning after Mackenzie had had Henderson and Wallace in his office. But it still did not streamline a scientific organization that was involved in numerous and complicated projects competing for time and manpower. Managing this was not in J. T. Henderson's line, either.

What happened in the fall of 1941 is only partly clear, but it appears that at about the end of November there was a "palace revolution"; a small delegation from the scientific staff waited on Boyle and told him that something must be done. They seem to have made the point that this "something" was not the displacement of Henderson by Boyle. Boyle took the matter to Mackenzie, who called Boyle and Wallace to his office on December 2, 1941. Mackenzie "intimated that as far as the business end was concerned I would want to keep a close control through Colonel Wallace and that from the scientific and technical end there probably was a need for a more definite internal organization."[11] But this was a particularly bad moment to rock the boat, in view of the rush to ship a sample of the GL IIIC radar[12] to England. They agreed to think about it and wait until this had been done. The set left Halifax on December 10. On the twenty-second Mackenzie "had a long talk with Colonel Wallace re organization of radio section."[13] On the thirty-first he spent another hour with Wallace and on January 2, 1942, he saw him at 2:20 P.M., and then at 5:30 had Henderson in his office to discuss the reorganization. On the fifth the whole afternoon was devoted to the matter, including a two-hour meeting of the senior members of the radio staff; "Had Dr. Henderson

[10] Mackenzie, "War Diary," August 18, 1941.
[11] Ibid., December 2, 1941.
[12] See p. 69 below.
[13] Mackenzie, "War Diary."

and the section heads in and I told them what we were prepared to do."[14] At 6:10 P.M. he saw Wallace again, and the next morning and afternoon met with Wallace, Bell, West, and Happe and worked out the details.

There is no mention of Boyle in these negotiations, but on January 9 Mackenzie met with Boyle and Wallace and formally decided to put the new organization into effect at once. The very next day saw the first meeting of the Co-ordinating and Management Committee that had been provided for, consisting of Wallace as Chairman, J. T. Henderson, and the heads of the Sections as noted above. From this moment the Radio Branch became *de facto* entirely independent of the Division of Physics, and I can find no further reference to Boyle.

The fact that Mackenzie devoted so much time in his extremely busy schedule to this matter shows how essential he believed such a reorganization to be. While there are no minutes of the meetings held in his office, it seems likely that the engineer-administrator and the solider-industrialist saw what had to be done and laid down the law to a group of bright and enthusiastic young men. At any rate the results were more than satisfactory.

One of the survivors believes that what finally led the "grievance committee" to call on Boyle was the circumstance that J. T. Henderson's secretary was a very strong-minded lady who made access to her "boss" extremely difficult and, they thought, was trying to run the Section herself. This sounds like the sort of frustration that causes revolutions; but it all ended well. Wallace ran the Branch with conspicuous success until at the end of the meeting of the Co-ordinating and Management Committee on Saturday, December 15, 1945, he suddenly announced his resignation, to the astonishment of everyone else at the meeting. At the same time a new Electrical Engineering and Radio Branch was formed, under the direction of B. G. Ballard.[15]

The role envisaged for John Henderson in the new organization is not clear from the minutes of the Co-ordinating and Management Committee. However, a letter written on January 15, 1942, by Colonel Wallace to K. A. MacKinnon, one of the members of the professional staff who was in Halifax at the time,[16] makes it seem quite definite:

> Dr. Henderson is a member of the Committee as chief scientific officer. His duties are to generally act as scientific and technical adviser to the Board for the co-ordinating of all scientific and technical activities of the laboratory Sections, and to act as liaison with all similar organizations in

---

[14] Ibid., January 5, 1942.
[15] The work of Ballard and the Electrical Engineering Section during the war is described in Middleton, *Physics at the National Research Council*, chap. 7.
[16] N.R.C. file 45-2-32.

the U.S. and the U.K., and to be responsible for gathering and dissemination of information in connection with the radio programme.

At the meeting of the Committee two days later it was agreed that Henderson would take monthly reports from the Section heads and assemble them into a monthly report from the Branch before the eleventh of each month—not an important job, one would think, for a "scientific and technical adviser." In fact, in this rapidly growing field, the extremely bright group that he had recruited evidently felt little need of such advice.

Henderson would have been more than human if he had not felt that he was being pushed aside, although others of the professional staff who have been interviewed agree that he made no complaint. It is clear that Mackenzie had a high opinion of him as a physicist, and when on April 5, 1942, Henderson told him that he had been approached by the Air Force to join them and go on operational research, the acting President told him that he thought he could be useful there, but "we would let him go only on the basis that he remained in a consulting capacity with us and considered himself still part of our organization."[17] Henderson did in fact take a commission in the Air Force with Mackenzie's consent, as the latter told McNaughton on July 14. "It will be his duty to see that the many RDF equipments, both airborne and on the ground, are thoroughly understood and used by the fighting forces."[18] He was given leave without pay on September 10, 1942, and returned to the Radio Branch on May 31, 1946.

The mention of K. A. MacKinnon demands a short note. MacKinnon had graduated in both physics and electrical engineering at Queen's, and had spent a few months in the Division of Physics before going to the Canadian Radio Broadcasting Commission in 1932. Thus he had been in the field of radio longer than anyone else in the Radio Section, to which he came in September 1940, and immediately became involved in work on antenna design for the ASV[19] and CD[20] radars. In January 1942 he was at the east coast conducting tests of the latter. "If you had been here," Wallace wrote in the letter referred to above, "you naturally would have been in the various meetings and consultations which took place." Later in the letter Wallace notes that the question of a special Antenna Section was raised but that it was decided to make this a part of the Special Research and Development Section; "but unfortunately we were deprived of the benefit of your advice . . . and if you have feelings that this should be reopened, the

---

[17] Ibid., April 9, 1942.
[18] Mackenzie-McNaughton Wartime Letters, p. 111.
[19] See p. 96.
[20] See p. 67.

Committee will be very glad to discuss the whole matter with you on your return." Wallace evidently had a high opinion of him, but he never became head of a Section and hence a member of the Committee.

6.4 The Field Station

The first mention of a field station that I have been able to find is a memorandum from J. T. Henderson to the President on the desirability of having an area of at least 250 acres "located at some distance from Ottawa."[21] This was on April 9, 1938, and although General McNaughton was quite certain that war was coming, it is not surprising that nothing was done. In the autumn of 1939 it became clear that some action was necessary, and it was taken, but on a much smaller scale; twenty acres near the Metcalfe Road south of Ottawa were leased from a farmer, J. B. Boyd. On this field Henderson and his small staff tested their first experimental radar sets. They had two tents and one or two small trailers to shelter them from the Ottawa winter. "The adverse working conditions and the difficulty of keeping apparatus running in cold weather," says the *War History* in masterly understatement, "impeded progress."[22] On January 11, 1940, Henderson outlined his requirements for a frame building, and after what seems now to have been a great deal of negotiation, a contract was awarded on March 9 in the sum of $10,571. For some obscure reason the architect of the Canadian Broadcasting Corporation in Montreal had to design this little building. But this was during the "phony war," and the careful habits acquired in the 1930s had not begun to evaporate.

The building was finished in May, and the need for another became apparent almost at once. The Department of National Defence undertook to pay for this, and it was finished in August 1941. A third building, assembled by joining together fourteen wooden huts, was completed and provided with central heating by the end of that year. Before that time a remarkably successful demonstration of the GL Mark IIIC radar had been held at the field station on July 23.[23] The Americans for whom it was put on were greatly impressed with the demonstration and astonished at the primitive facilities. To quote Mackenzie:

> ... the American group said that if their organizations had done as much and as satisfactory work as we had done in the last nine months they would consider the construction of a million-dollar building on our farm

---

[21] N.R.C. file M6-2-1.
[22] "War History," p. 10.
[23] See p. 76 below.

justified. Incidentally, we have two buildings on that site which cost fifty thousand dollars, and a number of tents and other shacks.[24]

Perhaps enthusiasm outweighs magnificence, at least in wartime.

The only other considerable structure built by the N.R.C. at the field station during the war was a timber tower, 200 feet high, put up to support the antennas of the LREW radar.[25] However, the Navy and the Air Force did provide their men with some acommodation in 1942. A view of the station, taken from the top of the tower shortly after the war, is shown in Figure 6.2.

The twenty-acre site itself was purchased early in 1941 for $50 an acre, and surrounded by a fence during the summer of that year. An additional eighty acres were rented next to it to accommodate measurements of radiation patterns.

Life at the field station was no picnic. Until some time in 1945 the road from the Metcalfe Highway was unpaved and often very difficult. The buildings were always overcrowded, and this is not surprising, for on January 12, 1942, the Co-ordination and Management Committee were told that ninety people were working at the station. It is recorded on the file that on December 13, 1941, the septic tank of the first building overflowed.[26] There is no indication of how they managed, but on another file[27] it is recorded that a new tank was not quite completed on August 17, 1942. People often worked long hours, and it must have been very helpful when a cafeteria began serving meals on July 28 of that year. During 1943 twenty "Wartime housing units" were built nearby to provide accommodation for twenty families of the workers.

The minutes of the Co-ordinating and Management Committee are liberally sprinkled with items relating to the field station; problems of security (real and imagined), allocation of space, provision of electrical power, and transportation between the station and the Sussex Street building. Few employees had cars of their own, and there was a great deal of coming and going by the professional and technical staff. Organized by the Red Cross, volunteer drivers, including many ladies from Ottawa's *beau monde*, operated N.R.C. vehicles on regular schedules to provide such a service.

6.5 Training of Service Personnel

The effectiveness of any instrument depends upon how skillfully it is used. It is natural that one of the ways in which the National Research

---

[24] *The Mackenzie-McNaughton Wartime Letters*, p. 85.
[25] See p. 98 below.
[26] N.R.C. file M6-30-1.
[27] N.R.C. file M6-30-13.

FIGURE 6.2
View of the field station

Council was asked to help the Armed Services was to train a small number of people to operate and maintain each of the numerous types of radar equipment, with the intention that these trained people should then instruct the much greater number of operators and maintenance staff needed in actual operations.

This training began in a very small way just as the war was starting in 1940. Three naval ratings were taught to operate and maintain the "Night watchman" set that was to be installed near Halifax.[28] In the spring of 1941, fifteen sub-lieutenants of the Royal Canadian Naval Volunteer Reserve (R.C.N.V.R.) were not only taught to operate the CSC set,[29] but probably became even more familiar with this apparatus by helping with the construction of several more of these sets for the use of the Navy.

During most of 1941 the main activity of the Branch was the development of the GL IIIC radar.[30] Even while the laboratory prototype was being put together, some soldiers were given instruction on its operation. As a large number of these sets were to be built, and the Army had not yet established a school of its own for training the maintenance men and operators, no less than eighty-eight men were given two months' instruction by the Branch, and a hundred more were trained as operators in courses lasting only a fortnight.

An interesting part of this programme of instruction was the training of eight young women—secretaries and stenographers in the Radio Branch—to operate the GL Mark IIIC set. During the late spring of 1941, when the prototype was being made ready for testing, a short course and a very brief practice period made these women into excellent operators, who were employed in the very successful demonstrations held during the summer for official visitors. This was an excellent public relations gesture in view of prevailing attitudes ("even girls can operate it"), but as a matter of fact women seem to be very good in such occupations, and after seeing this experiment the Joint Inspection Board of the United Kingdom and Canada decided to employ women for the final testing of the GL Mark IIIC sets in production at Research Enterprises, Ltd. These were members of the newly formed Canadian Women's Army Corps (C.W.A.C.), ten of whom were trained for this work at the National Research Council.

As the war progressed and the various kinds of radar equipment took shape in the Radio Branch, a certain number of people from one or more of the three Armed Services were always attached to the Branch, and this ensured that a nucleus of men who were familiar with the operation and maintenance of these radar sets would be available to the Services.

---

[28] See p. 47.
[29] See p. 48.
[30] See p. 69.

6.6 Radar and the Universities

The relation between the National Research Council and the Canadian universities has always been mutually beneficial. It has often been stated that without the help of the Council the establishment of postgraduate schools in the sciences would have been greatly hampered, and in most cases would not have taken place. On the other hand, the graduates of these schools have formed an indispensable source of scientific manpower for the laboratories of the Council.

At the beginning of the war the full employment of the universities in military research was inhibited by the necessity for secrecy. This was particularly so with radar, in particular microwave radar, for it was not known whether the enemy had developed this technique at all. In the United States it was decided in September 1940 to concentrate microwave research in one laboratory in Cambridge, Massachusetts. After the visit of the Tizard Mission the acting President of the N.R.C. became convinced that under proper conditions of security it would be beneficial to bring the Canadian universities into the picture.

This decision was almost certainly hastened by events that had been taking place at the University of Western Ontario. Professor R. C. Dearle of that institution had been told by R. W. Boyle that radio would be an important field in the war effort, so that in December 1939 the Department of Physics at Western "decided, as a war policy, to devote all of its resources in teaching and research to radio."[31] They were not at this time informed of the existence of radar, but they guessed correctly that microwaves would be important. They bought a low-power magnetron—not, of course, a cavity magnetron—from an American manufacturer and began to learn how to make it work.

In the summer of 1940 Dearle, or someone else at Western, still having no word of radar, sensed the problem and the direction in which its solution should be sought. The Department applied for an N.R.C. grant "to examine means for the development of a portable device which can be used to establish automatically the co-ordinates of a distant object,"[32] and suggested three essential steps towards that end. Whoever examined the application at the N.R.C. must have had a surprise. The workers at Western realized this later: "this modest offer to take over the whole of the microwave research of the United Nations," as the statement whimsically records, "brought an immediate response from N.R.C. and from British Liaison that we might be satisfied with some particular part in the development."

This response was given at a meeting held in Ottawa on November 19, 1940, with Mackenzie in the chair and Professor R. H. Fowler,

---

[31] Mimeographed statement, undated (ca. 1945) from the University of Western Ontario.
[32] Ibid.

the British Liaison Officer, at his side. Physicists from McGill, Queen's, Toronto, and Western Ontario Universities attended, together with Professor A. G. Shenstone, R. W. Boyle, representatives from the Army and the Air Force, and about a dozen people from the Radio Section, including, of course, J. T. Henderson. Fowler gave a report on the state of the art in Great Britain, and referred to the microwave experiments in the United States. Henderson gave a general account of what had been done at the National Research Council in Ottawa.

Then Professor J. S. Foster reported on microwave experiments underway at McGill. It is rather remarkable that even in the absence of information about the development of radar the people at McGill were also working in useful directions with microwaves. R. W. Mackay told the meeting that the Physics Department of the University of Toronto had been experimenting on superheterodyne receivers for microwaves. At Queen's they were working on methods of measurement of somewhat lower frequencies.

There is no record of any further joint university visits to Ottawa, and although there was obviously much liaison, the details are hard to find. The University of Western Ontario did prepare the above-mentioned mimeographed statement, so that it is easy to determine what was done there. For the first twenty months this was mainly the measurement of the radiation patterns of paraboloids of revolution and parabolic cylinders. During the winter of 1940-41 the receiving equipment "was mounted at the top of a low tower, with no protection from the weather," so that the researchers had to be rather heroic, but during 1941 a small frame building was erected. Methods of feeding radiation into a paraboloid were also studied. During the latter part of the war this group concentrated its attention on mixers for microwave superheterodyne receivers. At the peak of the work on radiation patterns, sixteen people at Western were engaged in this research.

The University of Toronto contributed to war research in a number of ways, but as far as radar is concerned its contribution was in the field of mathematical theory. Some members of the Departments of Physics, Mathematics, and Electrical Engineering formed a "Special Committee on Applied Mathematics"; the moving spirit was the Professor of Applied Mathematics, the extremely able J. L. Synge. The others whose names appear in the minutes of the C. and M. Committee are Colin Barnes of Physics and V. G. Smith of Electrical Engineering, both of whom were given all necessary secret information and came to Ottawa frequently. These men made mathematical analyses ad hoc of problems submitted to them by the engineers of the Branch.[33]

---

[33] Many of their results are dealt with in W. H. Watson, *Waveguide Transmission and Antenna Systems* (Oxford, 1947), chap. 10.

Queen's University "specialized in the development of techniques of microwave impedance measurements, matching measurements of dielectric properties, and design of resonant cavities."[34] That university, like the others, was lent necessary apparatus by the National Research Council.

Radar research at McGill University had an interesting career, which it is now probably impossible to document in detail. The two physicists most closely involved were a senior professor, J. S. Foster, and a brilliant younger man, W. H. Watson. On June 27, 1942, it was reported to the C. and M. Committee that Foster would "be given every facility at M.I.T. to keep himself informed on all branches of radio research there. Dr. Foster will then visit the National Research Council once or twice a month to advise the Radio Branch of developments at Boston." The acting President approved this. I have not found any reports of this activity. Foster's name is also attached to a scanning antenna that received further development after the war.

It is easier to appreciate the contributions of W. H. Watson, if only because no less than twenty-three of the reports in the PRA and PRB series issued by the Branch were written by him, often in collaboration with E. W. Guptill and J. W. Dodds. All but one of these deal with various aspects of microwave antenna design. His most brilliant invention was the use of a slotted waveguide as a radiator of microwaves, a concept that proved extremely useful; in September 1952 Watson and Guptill were granted two patents based on this idea.[35] Many of his papers show him as a very able mathematician.

It would appear that the radar work at McGill ended rather abruptly in the spring of 1944, for on May 27 we hear that "all secret equipment and machine tools at McGill will be returned to N.R.C."[36] This was because Watson left to become head of the Department of Mathematics at the University of Saskatchewan. On December 30, 1944, "It was agreed that Dr. W. H. Watson [as well as] the mathematical research workers at the University of Toronto may continue to request mathematical papers from the N.R.C. files bearing the rating 'Confidential.' No documents whatever are to be sent to McGill."[37] With Queen's and Western it was business as usual.

As a fitting conclusion we may quote a somewhat earlier minute of the Committee: "It was agreed that the results of the work being done at the Universities justifies the expenditure."[38]

[34] "War History," p. 16.
[35] Canadian Patents 486,637 and 486,638, issued September 16, 1952.
[36] C. and M. Committee minutes.
[37] Ibid.
[38] Ibid., February 19, 1944, minute 3.

6.7 Research Enterprises, Ltd.

The manufacture of large quantities of highly secret equipment presented two kinds of problem: first, of course, the requirement for extreme security, and secondly, the actual difficulty of finding industrial firms with the ability to met the requirements, even if the problem of security could be solved. The answer was the establishment of a large Crown company that could undertake all the work that could not safely be let out to sub-contractors. Such a company was Research Enterprises, Ltd.[39]

The first impulse towards the establishment of Research Enterprises, Ltd. came, not from the Radio Section, but from the concern that General McNaughton felt about the supply of military optical instruments. He remembered acute shortages of these during the First World War, and, knowing that a second war was inevitable, in August 1939 asked L. E. Howlett, head of the Optics Section, for a report on how an optical industry could be established in Canada. The immediate result of this was only the setting up of a workshop in the Optics Section for the fabrication of precise optical parts. But after the evacuation of Dunkirk at the end of May 1940, a floor plan for an optical factory was requested from Howlett, and provided a few days later. It was at about this point that J. T. Henderson associated himself with this project in the prescient belief that an organization would be required for the manufacture of radar apparatus. C. D. Howe, who had been made Minister of Munitions and Supply on April 9, 1940, saw the necessity for a Crown company. There is some doubt about the exact date of the formation of the company: Kennedy says August 16; an agreement with it was made by Order-in-Council on September 6,[40] but an advance of $50,000 had been granted on August 24,[41] and as early as August 13 G. W. Sweny, who had been appointed manager, had an office in the building in Sussex Drive.[42] What seems certain is that during August Sweny, Howlett, and Henderson were busy looking for a site for a factory. The vicinity of Ottawa was considered, but Sweny recognized that Toronto was a better choice because of the greater availability of labour. Finally fifty-five acres were found at Leaside near Toronto.[43]

The very first shortage at the beginning of any war is likely to be machine tools. Without proper financial authority, but doubtless with the blessing of C. D. Howe, the National Research Council ordered

---

[39] See J. de N. Kennedy, *History of the Department of Munitions and Supply, Canada, in the Second World War*, 2 vols. (Ottawa: King's Printer, 1950), vol. 1, pp. 407-41.

[40] P.C. 4664, 1940.

[41] P.C. 4147, 1940.

[42] Mackenzie, "War Diary," August 13, 1940.

[43] The purchase of this land was authorized only on November 2, 1940 (P.C. 6151).

about $600,000 worth from lists drawn up by Howlett and Henderson, certainly saving many months. The optical factory was begun on September 16, the radio factory about six weeks later. Meanwhile on August 27 Howe had appointed an able Ontario industrialist, Lieutenant-Colonel W. E. Phillips, to the board of the new firm. Phillips, whose managerial ability was a major factor in the success of the venture, replaced Sweny as General Manager on November 30, 1940.

In a very interesting and useful historical account, apparently written early in 1944,[44] we are told that an office was opened in Toronto on September 3, 1940. C. D. Howe pushed the new enterprise. In Mackenzie's diary we see that in the week of October 14 to 20 there was "much activity" by the minister, Phillips, Sweny, Hackbusch, and Henderson. This is his first mention of R. A. Hackbusch, who came from the Stromberg-Carlson Company and was made manager of the radio side of R.E.L. by Phillips, an appointment that turned out to be unfortunate.

On October 23 the members of the Tizard Mission[45] met officers of the Department of Munitions and Supply and discussed the production of radar equipment. On the twenty-seventh the scientific members of the Mission, namely, Cockcroft, Fowler, and Bowen, met Colonel Phillips in Toronto. On November 16 the Cabinet approved a cost-plus contract for a radar factory with an area of 50,000 square feet, together with a gatehouse and a cafeteria to serve 500 employees.[46] On December 28 another Order-in-Council authorized R.E.L. to make cathode-ray tubes and approved $750,000 for a building and equipment for this purpose.[47]

Phillips lost no time in establishing relations with the National Research Council. In his own words, "Our first step was to agree explicitly with the group at National Research Council that in this field they would function as our exclusive source of design information in every case where exclusive Canadian developments were concerned."[48] This statement calls to mind the fact, not generally known, that a great deal of the radar equipment produced at R.E.L. was copied, with slight modifications, from British designs, and much of this was supplied to the United States. The role of the Radio Branch in this was mainly to provide accommodation for groups of engineers from R.E.L. during the winter of 1940-41 while the factory at Leaside was being built. This building was completed in March 1941 and the machinery installed soon after. By August the company

[44] W. E. Phillips, "The Development of Research Enterprises, Ltd.," Typescript, 63 pp., appendix, charts, n.d. Public Archives of Canada, RG28A, vol. 17.
[45] See Chapter 5.
[46] P.C. 6552, 1940.
[47] P.C. 7665, 1940.
[48] Phillips, The Development, p. 31.

had radar orders in hand totalling $36,798,000.[49] To accommodate the heavy and bulky components of the GL radar[50] another building, with the appropriate cranes, was begun in July 1941, although the contract for it was not let until September 16.[51]

The relations between N.R.C. and R.E.L. were not always smooth. Each organization undoutedly had legitimate complaints about the other. One difficulty was that at R.E.L. the professionals were mainly engineers, but at N.R.C. most of the senior men were physicists. The sort of trouble that this could cause was well known to Mackenzie, who (albeit in reference to another programme) complained that "high-grade physicists sit around for hours discussing problems which are solved in the first chapter of elementary engineering texts."[52] Partly because of this, but even more because of very rapid technical progress and the urgency of the whole programme, an astonishing number of engineering changes were called for during the course of production. Phillips reported that from the beginning of the operation, presumably until the end of 1943, there had been 2,740 engineering changes imposed on the company, with a total added cost of $1,396,904.[53] Naturally most of these changes were blamed on the N.R.C.

On the other hand, the Radio Branch had serious complaints about R.E.L. These had to be dealt with carefully because of the peculiar triangular relationship between Mackenzie, Phillips, and the immensely powerful C. D. Howe. It is evident that Howe trusted Phillips entirely and that Mackenzie saw that he could not say anything to the Minister that could be interpreted as criticism of Phillips. At the end of 1941 R.E.L. were behind in their promised deliveries, and were trying to blame the N.R.C. On December 5 Mackenzie made an appointment with Howe at 3:30 P.M. His entry in the diary is illuminating:

> Discussed the matter of correspondence with R.E.L. and told Howe that we did not care to enter any paper war with R.E.L., that they were most optimistic in promises of delivery, were finding themselves in difficulties and apportioning blame which we did not feel extended to us. I indicated that we had a high regard for both R.E.L. and Colonel Phillips and thought that ultimately things would iron out. We were only concerned with how he himself felt about the situation. If he was satisfied that everything was all right we would not bother answering letters. He agreed completely, with very complimentary remarks about the work of the Research Council and said there was a general deterioration of the

---

[49] Ibid., p. 32.
[50] See p. 69 below.
[51] P.C. 7211, 1941.
[52] Mackenzie, "War Diary," November 9, 1942.
[53] Phillips, The Development, p. 54.

nerves these days, people were tired and there was a lot of loss of control. I told him we did not intend to do anything about it, but any time he wished to check up we had facts that would be satisfactory. It was a very interesting interview.[54]

Difficulties increased during 1942, and Mackenzie quickly became suspicious of the ability and actions of Hackbusch, who at first had the complete support of Colonel Phillips. It seems that quality control at R.E.L. was far from adequate. On November 11, 1942, Colonel Wallace went to Toronto with two of his Section heads "to find out what was actually wrong with R.E.L. production."[55] Wallace suspected that components were being put through without inspection in spite of denials by Hackbusch and his subordinates. It emerged that Colonel Phillips had not been aware of what was going on, and that Wallace's suspicions had been justified. A plant superintendent admitted that he had been told by Hackbush to concentrate on volume rather than quality. Mackenzie unburdened himself to General McNaughton on January 2, 1943:

> I am afraid that Colonel Phillips has not had the best help in his assistants as Hackbusch . . . is a grand buck-passer and is continually blaming everybody, including ourselves, the services, the Inspection Board, and in fact everyone he comes in contact with for the faults which are really his own. However, the Council has been particularly careful to avoid any criticism at all and has maintained excellent arrangements with Phillips and most of his crowd . . . we are leaning over backwards to prevent any friction while there is any possibility of improving matters.[56]

Wallace's November visit must have shaken Phillips' confidence in Hackbusch, and in the following March, on a visit to Ottawa, Phillips met Mackenzie and Wallace and "[admitted] all the weaknesses of Hackbusch and [said] they are going to make a fundamental change which we all know is overdue by at least two years."[57] But it was September 2, 1943, before Hackbusch left the firm. A week later Wallace was made Director of the Radio Division at R.E.L., but he also continued to direct the Radio Branch at the N.R.C. until December 15, 1945.

Phillips let Hackbusch down lightly:

> The problem in the Radio Division is especially difficult because of the complicated nature of the devices involved. The rapid expansion of the operation involved a high degree of decentralization and control.
>
> It became apparent at the beginning of 1943 that these conditions were not being maintained. Mr. Hackbusch, who had performed valu-

---

[54] Mackenzie, "War Diary," December 5, 1941.

[55] Ibid., November 12, 1942.

[56] *The Mackenzie-McNaughton Wartime Letters*, p. 123.

[57] Mackenzie, "War Diary," March 25, 1943.

able service in gathering his organization together, was obviously suffering from continuous overwork, and in September, 1943, we were forced to accept his resignation.[58]

Some of the sad results of this mismanagement will appear in Chapter 8. The situation was undoubtedly exacerbated by the presence at R.E.L. of officers of the Inspection Board from the United Kingdom who seem to have had no reputation for tact. Wallace quickly put things to rights, and when Mackenzie visited R.E.L. on March 30, 1944, he was enthusiastic. "The plant is a far different organization than the last time I saw it."[59] The path from prototype to production model had become smoother; Wallace was able to get co-operation all down the line. Several engineers from the N.R.C. were also sent to R.E.L. after Hackbusch left.

Research Enterprises, Ltd. went out of business in September 1946, after having shipped over $220 million worth of wartime equipment. At this time, the National Research Council acquired its field station on the shore of Lake Ontario at Scarborough.

6.8 Plan of Subsequent Chapters

The next three chapters of this book will deal with projects of research and development entered into by the Radio Branch during the war. These were, of course, undertaken at the instance of one or other of the Armed Services, and sometimes, after proper consultation, those of Great Britain. It is reasonable to begin with the Navy, not only because in the British Empire it was always considered the senior service, but also because the first radar to be put into actual operational use in Canada was constructed at the request of the Royal Canadian Navy.

In each chapter the various projects will be dealt with in the approximate order of their inception. To save space, each project will usually be identified in the text by an abbreviation. All these and other abbreviations used in this book will be found in Appendix A.

The indispensable track through the jungle of information about these wartime developments is provided by the official "War History of the Radio Branch,"[60] supplemented for the period before January 1, 1942, by a progress report written by J. T. Henderson.[61] The larger "War History" is almost completely anonymous,[62] the "Progess Report" somewhat less so. Fortunately some flesh can be put on these

---

[58] Phillips, The Development, pp. 58-59.

[59] Mackenzie, "War Diary," March 30, 1944.

[60] See footnote 4 of this chapter.

[61] National Research Council of Canada, Radio Section, "Progress Report for Period June, 1939 to 1 January, 1942," mimeographed, 10 illustrations.

[62] The only names I have found—all in the introduction—are those of Fowler, Heakes, and Henderson.

dry bones by the use of the minutes of the weekly meetings of heads of groups, already referred to, held almost every Saturday beginning on August 16, 1941, and on January 10, 1942, replaced by a Co-ordinating and Management Committee that continued to meet on Saturdays to the end of 1946. The physical progress of most of the projects can be followed from monthly reports beginning with January 1942.

For technical details there are two series of reports, designated PRA and PRB, issued by the Radio Branch.[63] PRA reports were intended to refer to work of sufficient interest to warrant their distribution to appropriate institutions outside the National Research Council, subject to the requirements of secrecy. PRB reports were considered to be of less general scientific interest and were not generally distributed outside the Council, although such distribution was not prohibited. For the purposes of this book it has been necessary to use these only occasionally.

---

[63] These are listed in Appendix C.

Chapter 7

PROJECTS FOR THE NAVY

The projects with naval applications will be dealt with in approximately the order of their inception. In addition, it should be noted that the Coast-Defence Radar (CD) described in Chapter 8, although developed at the request of the Army, was first used by the Navy.

7.1 Shore-Based Search Radar, "Night Watchman" (NW)

In wartime it is clearly essential to be sure that no enemy vessel, either surface or submarine, can enter a harbour without being detected. The entrance to Halifax Harbour was defended by a loop of wire arranged so that the magnetic properties of the ship would activate an indicator. In March 1940 the R.C.N. asked the Section to devise a simple form of radar to supplement this system.

This was the simplest possible application of radar, as it required merely a fixed and not necessarily very narrow beam of radiation, directed appropriately across the channel. The transmitter, operating on a wavelength of 140 cm., had a power output of about 1 kilowatt and a pulse-length of about 0.5 microsecond. Two antennas were used, one for the transmitter and the other for the receiver. By the middle of June 1940 the set was operating well in the laboratory, and was installed at Herring Cove near Halifax in July. Three navy ratings were instructed in the operation of this radar set, which performed very well, besides giving the staff of the Section experience in high-frequency engineering. An improved receiver was installed in the spring of 1941.

It is interesting that the U.S. Army Signal Corps requested a similar set, which was made for them in about two months.

The reader may wonder why this apparatus was called "Night watchman." It seems that this was because, having no automatic alarm, it had to be watched all night, and in foggy weather. On a clear day it would not be needed.

7.2 Ship-Borne Radar on Metre Wavelengths (CSC, SW-1C, SW-2C, SW-3C)

An obvious naval application of radar is to enable the commander of a ship to detect other vessels and features of the coastline in darkness or fog. Indeed, during the first decades after the invention of radio, such an application was envisaged by various people, long before the existence of the techniques for making it. I suppose that the post-war generation of mariners may be unaware of the sense of groping isolation that their predecessors felt in dense fog, the eerie silence on the bridge as the ship edged along at quarter-speed, all ears strained for the sound of a ship's siren or the foghorn on some promontory, unreliable guides because of the peculiar acoustic properties of fogs.

In February 1941 the R.C.N. asked the Radio Section about the possibility of developing a radar apparatus to be installed in naval ships for the detection of surfaced submarines. On March 19 formal arrangements were made for this. It was suggested at first that the ASV[1] transmitter and receiver should be used, as it had already been developed, but it was decided to build an equipment especially for this purpose, using a wavelength of 1.5 metres (200 MHz). The skill and enthusiasm of the staff were so great that on Monday, May 12— fifty-four days later—sea trials were begun, with the apparatus installed in the corvette H.M.C.S. *Chambly*, near Halifax. The first day they just looked around and were happy with the indications they obtained. The second day's tests are worth reproducing:

> *Tuesday 13 May.* The weather was very foggy and visiblity about 1/4 mile, nevertheless ... a corvette and a Dutch submarine accompanied us to sea for routine tests. The other corvette stood guard over the submarine at a distance of about 1 mile while we steamed away from them. The submarine gave an indication up to 2.7 miles, and we continued on for an additional mile, turned and the Captain [i.e., the commander of H.M.C.S. *Chambly*] headed in the general direction of the submarine. At 2.7 miles we again detected the submarine and gave bearings to the Captain. He insisted that the submarine should be 28° from the bearing we gave him. Nevertheless we continued by RDF bearings giving the range to the Captain at intervals of 1/10 of a mile.
>
> We sighted the submarine absolutely dead ahead at 1/4-mile range. The Captain was delighted with these results, and we repeated the tests for him a second and third time.
>
> During these operations we reported ships at five miles and kept a record of their range and bearing and in every case we approached them

---

[1] See p. 96 below.

through the fog with such accuracy that collision contact would have been made if we had continued on the bearing.[2]

But at the quarter-mile range they sheered off to starboard. One can imagine the commander of H.M.C.S. *Chambly* grudgingly deciding that the young men from Ottawa with their black boxes really had something. I have been in some of these fogs off Halifax—albeit in peacetime—before the advent of radar, and I can imagine the impression that this made.

On May 16 more extensive tests were conducted, simulating an attack on a convoy. The naval authorities were completely convinced.

The radio engineer who was looking after this project was H. Ross Smyth, who was in charge of radar projects for the navy throughout the war.

The success of the tests led to an urgent request for a number of such radar sets. The quickest way to get them was to build them in the Radio Branch. To help in this, the R.C.N. attached fifteen sub-lieutenants to the N.R.C., and with their assistance seven more equipments were delivered to the navy. This was a stop-gap procedure and ceased as soon as R.E.L. had an adequate specification for the sets to be produced in quantity. At this time the apparatus was given a new designation, SW-1C, and later SW-2C after it had been modified so as to operate on a frequency of 215 MHz, in order to make it possible to use it in conjunction with existing equipment for the identification of friendly aircraft and ships (IFF).[3]

Several hundred sets were produced at R.E.L., and these remained in service until they were replaced by the more efficient microwave radar equipment RX/C.

We shall not go deeply into the technical details of these radar sets. The transmitter was intended to be a copy of the British ASV transmitter[4] except for the use of Canadian tubes and components; the receiver a simpler version of the one used on NW. The main design problem concerned the antenna. A rotatable Yagi antenna was adopted.[5] This was rotated by a large handwheel in the radar cabin through a flexible shaft leading up the mast (Figure 7.1).

For a similar set to be used in motor torpedo boats, requested by the R.C.N. in December 1942, the antenna was to be rotated by an electric motor at the masthead, controlled from the cabin. However, experiments had begun in September 1941 on electrical means of rotating an antenna at the masthead. Weight was the problem. There

---

[2] N.R.C. file B.4-R.4-5.

[3] "Identification, friend or foe." See p. 78.

[4] See Chapter 9.

[5] The Yagi antenna consists of one active dipole with one reflector dipole behind it and a number of director dipoles in front. It is a familiar type of television antenna in North America.

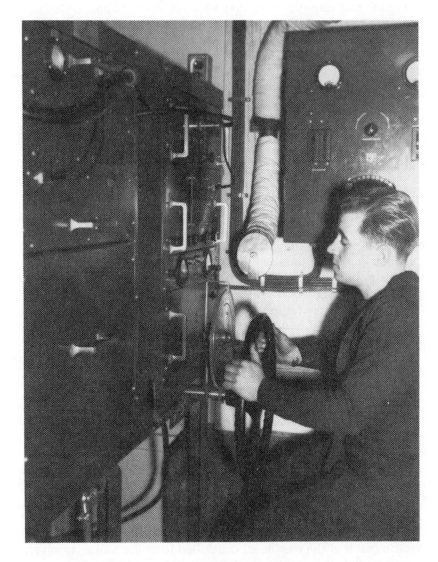

FIGURE 7.1
Operator at console of SW2-C
(Public Archives Canada, PA 105667)

was also much discussion of the means of feeding the antenna; open wires had been tried, but these were finally replaced by a coaxial cable.

7.3 Microwave Ship-Borne Radar (RX/C, SS-2C)

The disclosure of the cavity magnetron by the British, and subsequent experiments both in the United Kingdom and the United States, made it likely that for naval purposes, especially in small vessels, a radar set working on a frequency of 3000 MHz (a wavelength of 10 cm.) or thereabouts would have great advantages with respect to size and weight, and probably accuracy.[6] As a microwave radar was already under development for the R.C.A.F. (see Chapter 9), the opportunity of testing this hypothesis came in August 1941, when an experimental S-band radar was tested in comparison with the SW-1C at Duncan's Cove, just outside the entrance to Halifax harbour. The microwave radar provided greater range and better discrimination between targets on the surface of the sea.

In September, after a discussion between the Radio Section and Naval Staff, the Section began the development of such an equipment, while still improving the set that used the 200 MHz band. The project bore the N.R.C. designation SS-2C, and was later known by the R.C.N. code RX/C. The discussion with the Navy laid down the main characteristics desired: the transmitter and receiver, mounted in separate cabinets (Figure 7.2), would operate with a common antenna, rotated either by a motor or by a handwheel and flexible cable, as in the set illustrated.

It will be understood that the development of such a radar set involves a number of interlocking problems. Given the magnetron oscillator, a pulser has to be constructed to provide it with powerful pulses of voltage, a method of feeding this power to the antenna has to be chosen, and switches have to be developed to isolate the receiver when the pulses are going out (see Chapter 2). A suitable antenna has to be designed. The receiver must be adequately sensitive and its output must be fed to an appropriate cathode-ray-tube display, often several such displays. Electro-mechanical means of converting the signals on the cathode-ray tube to accurate ranges have to be provided.

The availability of large amounts of such high-frequency energy led to—indeed demanded—a revolution in circuit design. The wiring usual in electrical circuits was replaced by waveguides, typically made of rectangular metal tubing with the greater dimension of the

---

[6] The region of the electromagnetic spectrum in the vicinity of 3000 MHz came to be called "the S band."

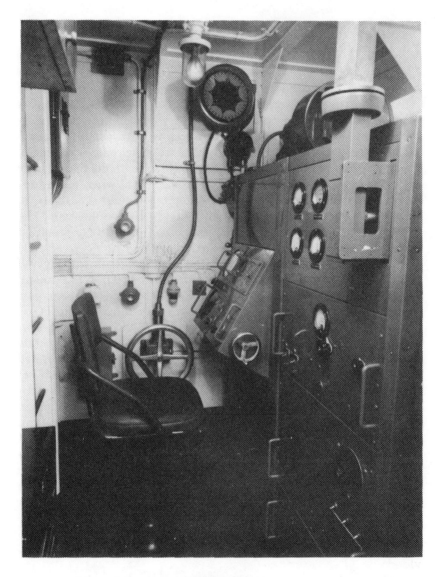

FIGURE 7.2
Installation of RX/C
(Public Archives Canada, PA 123628)

interior cross-section slightly more than half a wavelength. Bends in such guides, and the connections between them and the various components of the apparatus, had to be designed and made with great care in order to avoid serious attenuation of the high-frequency power. Because of the appearance of these circuits they became known as "microwave plumbing."

It is a fortunate circumstance that the techniques used in one kind of radar apparatus can generally be adapted for use in another. This fact, combined with the co-operation between the British, Americans, and Canadians, made possible the rapid development of this and other radar sets. In the present instance a great deal of the knowledge that had been accumulated in designing the GL Mark IIIC set (see Chapter 8) was very useful.

Nevertheless the RX/C radar was gradually altered and improved in the years 1942 to 1944. In 1942, at the request of the Navy, the equipment was physically redesigned so that it could be installed in a small hut on the bridge of the ship. This also made possible the substitution of a waveguide for the coaxial cable that led the power to the antenna. A plan position indicator (PPI) was felt to e a valuable feature; also a long-range scale for use with this as an aid in identifying coastlines up to 75,000 yards. These modifications were made in 1943 and incorporated in the production models coming from R.E.L., not, of course, without delays and frustration on the part of both R.E.L. and the Radio Branch.

In January 1943 the Navy was so impatient for these microwave radars that they asked the Branch to build eight RX/C equipments as quickly as possible, with PPI units to follow later. These sets were completed by July 1943. Various further modifications and improvements followed, and the RX/C radar saw service until the end of the war. Not without having their troubles; on February 11, 1944, the Secretary of the Naval Board wrote to Wallace enclosing a report from Commander-in-Chief Canadian North Atlantic complaining about the difficulty of servicing these sets. In April the Navy asked for an N.R.C. representative at sea trials to investigate this, and on May 1 Wallace detailed W. M. Cameron to attend the trials. It transpired that part of the trouble was due to careless handling and maintenance by naval personnel.[7]

7.4 Ship-Borne Radar Type 268 (RX/F)

The success of 10 cm. radar rather naturally led to the expectation that even shorter wavelengths would provide further advantages in compactness and accuracy. By the spring of 1941 it had been found

---

[7] The correspondence is in the Public Archives, N.R.C. file 45.2.52.

possible in both the United Kingdom and the United States to make a cavity magnetron to produce 3-cm. waves, or a frequency of 10,000 MHz.[8] On October 30, 1941, L. E. Howlett, then N.R.C. Liaison Officer in London, wrote to the acting President that he had been asked by the Director of Scientific Research, Admiralty, to find out whether such magnetrons could be produced in Canada.[9] The reaction to this was to wonder why 3-cm. radar sets could not be produced as well. A project of this sort was mentioned in a tentative research programme for 1942 laid out at the first meeting of the heads of groups in that year, held on January 3. Later the Microwave Section was expanded to include a group studying the techniques of 3-cm. radar.

By May 1942 this group had progressed to such an extent that when a formal request for the development of a high-resolution radar for small craft was received from the British Admiralty, the Radio Branch at once put this project in hand. A meeting of representatives of the Admiralty, the R.C.N., and the N.R.C. was held at this time and discussed the requirements for such an equipment.

These indeed presented a formidable problem. The set must perform at least as well as the larger ones then fitted in naval vesels, and at the same time it must be compact and relatively light.

Work proceeded concurrently on four aspects of the problem: the radio-frequency components, the rotating antenna, a plan position indicator,[10] and a compact modulator. By September 1942 these components had been developed to such a state that a preliminary model could be tested near Halifax in October, with gratifying results, in particular as to range. The success of these tests was reported both to the British and the Americans.[11] From this time on the project became a matter of engineering. It was necessary to minimize the weight, size, and power requirements of the apparatus because of the small size of the vessels in which they were to be installed. The equipment must be very rugged to withstand shocks from waves and gunfire, and waterproof as well. Because the Admiralty wanted to use these sets in the tropics, they had to be able to withstand long periods of high temperature and humidity. Finally there was no chance of making repairs at sea, so that great reliability was required.

Research Enterprises, Ltd. were brought into the picture at this stage. On November 27, 1942, K. C. Mann went to Toronto, armed with drawings and specifications, and met the engineers at R.E.L. The latter did not think much of the mechanical details of the prototype, and argued that the mechanical design should be left entirely to them.

---

[8] This region of the spectrum is called the X band.
[9] Public Archives, N.R.C. file 45.2.59.
[10] See p. 5.
[11] Letters and cables in Public Archives, Ottawa, N.R.C. file 45.2.59, October 1942.

Mann, always a master of conciliation, assured them that the mechanical details included in the N.R.C. specification "applied only to the N.R.C. model and were intended only as guides, not working specifications for the use of R.E.L."[12]

The project went slowly at R.E.L., far too slowly to suit the Navy. On February 8, 1943, Colonel Wallace reported to the C. and M. Committee that Lieutenant-Commander T. J. Brown had offered to furnish fifty radio mechanics to the Radio Branch to build some sets for the R.C.N. As machine tools and supplies were not adequate for this at Ottawa, it was suggested that Wallace and Smyth try to get R.E.L. to give these sets special priority, or else to furnish critical parts so that they could be made at the N.R.C. Conversations at Toronto between the Navy, the N.R.C., and R.E.L. allowed Wallace to announce a wek later that R.E.L. had agreed to construct ten hand-made sets by July 1943. The C. and M. Committee, evidently doubting that this would be done, agreed that "three or four" sets would also be made at the Radio Branch. In fact, the Branch eventually built eight of them. Hackbusch at R.E.L. heard about this and was far from pleased.

It is interesting that K. C. Mann was asked at this time to go to Washington as Scientific Liaison Officer. Fortunately he resisted this appointment, for by April a 268 set was furnished and sent by air to England, and on May 5 Mann arrived in London. H. W. McCrae and J. G. Retallack had to follow by sea. The equipment was installed in a motor torpedo boat at Portsmouth, much to the annoyance of a young red-haired commander who was incensed at these Colonials drilling holes in his vessel. The results of the trials soon turned his rancour into enthusiasm, and Mann remembers that this officer wanted to go across the Channel at once and have a crack at Jerry, but was dissuaded by the argument that the apparatus might fall into enemy hands.[13]

The tests of the 268 set were so successful that on July 15, 1943, Mann informed Wallace by cable that the Admiralty wanted to order 1,000 sets.[14] Actually a contract was entered into for 1,500 sets and 1,500 sets of spare parts.[15]

The sets that were ordered, however, were not identical with the one demonstrated by Mann, which had met all the requirements of the Admiralty except that it had a non-linear display in the first 1,000 yards of range. This had resulted from the use of a rotary spark gap as a modulator or pulser, adopted in order to save space. Fortunately a type of fixed and triggerable spark gap had just become available,

---

[12] Ibid.
[13] Conversation with K. C. Mann, November 10, 1978.
[14] Public Archives, N.R.C. file 45.2.59.
[15] Ibid., s.d. August 27, 1943.

ensuring satisfactory linearity without increasing the size of the equipment.

The appearance of the 268 radar eventually produced is shown in Figure 7.3, which also shows a separate PPI display unit which was to be mounted elsewhere in the ship. The rectangular tube extending out of the top of the picture is the waveguide that feeds the power to the rotating antenna, shown in Figure 7.4.

This excellent apparatus, probably the most distinctive Canadian contribution to the field of radar, was not arrived at without much struggle and delay, some of it certainly caused by conflicts of personality, although nothing of this sort appears in the official "War History of the Radio Branch." To get things moving,

> It was decided that the necessary development would continue at Ottawa while production design would go forward at Research Enterprises Limited under the guidance of a group of Radio Branch engineers. A policy requiring the separation of closely related design and development programs is not normally desirable but in this instance, largely due to continuous travel back and forth between Toronto and Ottawa by the Radio Branch engineers, it was thoroughly successful. Upon completion of the development work at Ottawa, this engineering group remained at Toronto until the production order was completed just before the end of the war.[16]

It ought to be remarked that many of the "engineers" were, in fact, physicists. This was true of K. C. Mann, who was appointed production manager by Wallace for the 268 radar at R.E.L. and remained in this post for most of the rest of the war. His remarkable talent for diplomacy certainly contributed just as much to the project as his technical abilities.

Going back to October 1943, a firm contract and rigid specification was drawn up by representatives of the Admiralty, the N.R.C., and R.E.L. The waveguide and the primary power source were to be supplied by the Admiralty. Requirements for anti-jamming features and prevention of interference with ships' radio facilities were written into the contract.

The "sea" trials of the engineered prototype were held in the presence of representatives of the Admiralty on Lake Ontario in July 1944. They were eminently successful, but for some reason that I have not been able to discover, full-scale production began only in December 1944, reaching 1,600 sets by the end of the war. Some of the first sets were sent to Africa for ships destined for the Far East, and W. L. Haney went to Egypt and J. Whealey to South Africa to superintend the fitting. There is an interesting correspondence on file

---

[16] "War History," p. 32.

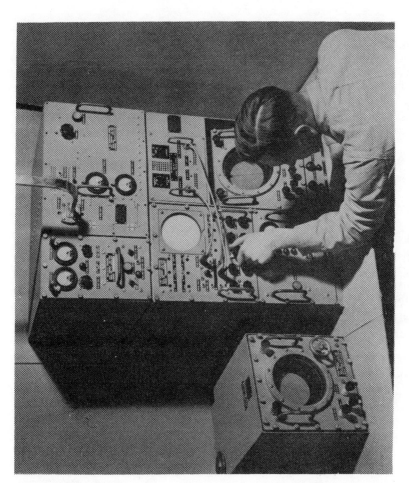

FIGURE 7.3
Type 268 radar, operator's console

FIGURE 7.4
Antenna of type 268 radar

45.2.59[17] about this. Writing to Wallace from London on April 20, 1945, Haney reported on the success of his mission and on some of the troubles discovered, partly the result of poor packing. He was concerned, however, that the Admiralty Signals Establishment seemed to be itching to make modifications to the sets sent to England. Brigadier Wallace replied on May 7 that he was disturbed by this:

> ... we have had many unfortunate experiences in the past due to various research groups in England being beset with a great urge to modify our equipment.... I must point out from the R.E.L. point of view that our production has now reached an output level of three hundred sets or more a month and any introduction of ... modifications at the present time would have a most disastrous effect on the flow of the sets and the morale of the workers.... Surely some of the operational personnel can persuade A.S.E. that the set fully meets the requirements and should not be tampered with.

It appears that some twenty of the sets were modified by the substitution of larger antennas, giving greater gain and a narrower beam, to enable the "schnorkel' or breathing tube of a submarine to be detected when the vessel was running submerged. Some internal circuit changes were also made, not all of which were certainly necessary.[18]

The 268 radar was the basis of a highly successful equipment for merchant ships, made in large numbers in Canada after the war, in spite of unscrupulous denigration by a competitor.

7.5 Cathode-Ray Direction Finder (CRDF)

As mentioned in Chapter 1, the main activity of J. T. Henderson in the years just before the war was in the development of the so-called "cathode-ray direction finder." This unfortunate name was adopted because the indicating device, which distinguished it from previous methods of direction-finding by radio, was the "cathode-ray[19] tube" or oscilloscope, now familiar in the form of the picture tube in a television receiver, but then strictly a piece of laboratory apparatus.

The CRDF has three receivers connected to three aerials having different directional responses. The output of these receivers, which have to have identical gain and phase-shift, is applied to the deflecting electrodes of an oscilloscope in such a way that a radial straight line appears on the screen in the direction of the distant transmitter,

---

[17] In the Public Archives.

[18] Personal communication from W. L. Haney.

[19] The term "cathode-ray" was coined by Hittert in 1876 for the stream of particles, later called electrons, that moves away from the cathode when an electric field is established in an evacuated space.

the actual position of which has to be determined by graphical computation from bearings taken at two or more CRDF stations.

It should be emphasized that this technique has nothing whatever to do with radar. Its usefulness in wartime is greatly reduced by the necessity of keeping radio silence, but it has an important application in helping ships or aircraft which have lost their way.

A short-wave CRDF set operating in the frequency range 1.5 to 7.2 MHz was under development in the Radio Section when war broke out. H. R. Smyth was on the east coast during the summer of 1939, setting out equipment at Portuguese Cove, Nova Scotia.[20] Field trials were held in the winter and spring of 1940, after which the equipment was put in service for one year at Halifax for the Department of Transport.

In October 1941 J. T. Henderson was informed that the British were looking for a Canadian firm to make 200 short-wave CRDF sets. At about the same time,[21] the R.C.A.F. asked for one to be installed at Dartmouth, Nova Scotia. It is not clear how things progressed during the following year, except that the project was under the direction of H. R. Smyth, with G. A. Miller in charge of the design of the antennas. But a prototype equipment was installed in a field near London, Ontario, and in February 1943 the design was approved by the Services. Orders for about a hundred complete sets were placed with the Sparton Radio Company during the spring of that year. The first antennas were delivered to the Navy in September, and the first receivers in March 1944. Engineers from the Radio Branch gave much assistance to the company in this work, and the Branch conducted field trials at the first naval installation.

Four models were produced, three of which worked in different bands of the short-wave radio spectrum, while the fourth was capable of covering the entire region between 2.7 and 25.0 MHz. It is stated in the "War History" that "the production model was the most sensitive visual DF used during the war."[22]

7.6 Observations of the Ionosphere

The term "ionosphere" is applied to the region in the upper atmosphere in which there occur one or more diffuse layers having a high concentration of free electrons. These electrons are formed by the ionization of molecules of atmospheric gases by ultraviolet light from the sun. The existence and variations of these layers are of very great

---

[20] N.R.C. file 45.2.14. Ross Smyth to R. W. Boyle, June 3, 1939.

[21] The "War History" places this request in December 1941, but the minutes of a meeting of heads of groups record it on October 25.

[22] "War History," p. 35.

importance to short-wave radio, for the following reason: the signals received at considerable distances do not go from the transmitter along the ground to the receiver, but up into the atmosphere until they reach a layer of free electrons, by which they are "reflected"—really sharply bent—downwards toward the ground. In any given situation there is a maximum inclination for total reflection (see Figure 7.5).[23] More steeply inclined rays will be reflected very little. Thus there will be almost no reflected wave from the transmitter T that strikes the ground nearer to it than B. The ground wave TB″ is rapidly attenuated, so that there may often be a "skip zone" between B″ and B in which neither the sky wave nor the ground wave can be received. The maximum angle for total reflection varies with the frequency, and this is why a knowledge of the electron layers is so important.

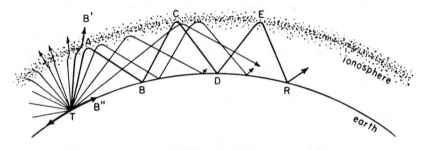

FIGURE 7.5
Total reflection of radio waves

The construction of equipment for measuring the height and other characteristics of the ionospheric layers was given a relatively low priority in comparison with the developments in radar. Nevertheless, in October 1941 the Navy put in an urgent request for a manually operated apparatus to work over a frequency range of 1 to 14 MHz and simply measure the heights of the layers. This apparatus was designated P′F and was completed and installed at Chelsea, Quebec at the end of the year, remaining in service throughout the war. It consisted of a manually tuned transmitter sending 100-microsecond pulses with a peak power of about 1 kW., and a commercial communications receiver. It appears that C. W. McLeish was in charge of this design, although there is very little reference to the project in the minutes of the weekly meetings.

In March 1943 the R.C.N. decided to extend the programme and requested the Radio Branch to build three more equipments incorpo-

---

[23] The figure is from John C. Johnson, *Physical Meteorology* (Boston: M.I.T. Press, 1954), by permission.

rating several improvements. This was called P'F Mark II. These three units had all been completed by August and were installed at widely separated places in northern Canada. In July 1944 the Branch was asked to modify these sets so that they would provide information on the density of the layers.

The interest in the ionosphere, and indeed in the general subject of the propagation of radio waves, continued to increase, and led to the formation of a Canadian Radio-Wave Propagation Committee. On the recommendation of this body the Radio Branch began in January 1945 to build an experimental automatic ionosphere recorder, designated AIR. An entry in the minutes of the Co-ordinating and Management Committee on March 31, 1945, indicated that it was to have the lowest priority of all the current projects, but the equipment was nevertheless finished in time to be set up at a station in Manitoba to observe a total solar eclipse on July 9, when valuable results were obtained.

This equipment had a transmitter and receiver that were simultaneously swept mechanically over the frequency range 1 to 16 MHz, together with a photographic camera to make automatic records of the oscillograph screen.

At the instance of the same committee, and in co-operation with a corresponding group in the United States, the Branch began in January 1945 to study the propagation of 3.2-cm. and 10.7-cm. waves over land. It was fortunate that an especially flat piece of ground was available at the research station of the Department of National Defence near Suffield, Alberta. This was large enough to permit a distance of 43 kilometres between two 25-metre towers, on the top of one of which the transmitters were placed, while the receivers could be placed at any desired height on the other. The path was almost ideal: flat to plus or minus 6 metres, crossing no water, and free from trees or buildings. The Department of Transport collaborated in this project by sending three trained meteorologists and the necessary instruments to make the elaborate meteorological observations required. It was late summer before the arrangements were completed at Suffield, but valuable observations were then obtained which were later studied in detail.[24]

7.7 Canadian Naval Jammer (CNJ)

The war was marked by a large number of uses of physics, and especially of radio, both offensive and defensive, by both sides. In the fall of 1943 a new German offensive weapon became unpleasantly effective in the Mediterranean. This was the radio-controlled glider

---

[24] See p. 120 below.

bomb, launched from high-flying aircraft and controlled by radio signals. But there was a defence. It was "found possible to make the enemy lose control of the bomb by jamming the control frequencies with a high-power frequency-modulated transmitter."[25]

Such transmitters were the object of a crash programme started at the request of the Royal Canadian Navy in February 1944. Five transmitters were requested, and also parts for twenty more to be assembled by the R.C.N. at St. Hyacinthe, Quebec. It is difficult to believe that the first equipment, complete with spares, was shipped to Halifax on March 28, and the other four in April.

The transmitter gave an output of 1 kW frequency-modulated at 150 Hz over 4 MHz in the region between 42 and 75 MHz. The output was also amplitude-modulated at 60 Hz; this may or may not have made it more effective but it certainly made the apparatus simpler by doing without a rectifier in the high-voltage supply. The frequency modulation was performed by a variable condenser rotating at high speed.

This project has only one oblique reference in the minutes of the C. and M. Committee. The description of the equipment was written by W. C. Wilkinson and R. S. Rettie. It was of no great technical sophistication and is interesting mainly as an example of the speed at which the Branch could respond to emergencies.

7.8 Radar Type 931

As the war progressed, microwave research made available shorter and shorter waves, that is to say, higher and higher frequencies, until in 1943 it was found possible to use a wavelength of about 1.25 cm., corresponding to a frequency of 24,000 MHz. This part of the microwave spectrum was designated the K-band.

We have already emphasized that the use of a higher frequency permits a more accurate determination of the direction of a target. It also allows increased resolution when two or more neighbouring targets are observed at once. A particularly valuable application of this, of great importance in naval warfare, is the ability to range on the splashes when shells, having missed their target, fall in the sea, so that errors in range and bearing can be corrected in the next round.

The Americans and the British were both working along these lines in 1943, and in December of that year representatives of the Admiralty came to Ottawa and asked the National Research Council to carry out similar developments in Canada. In particular the Radio Branch was asked to develop an antenna having a very narrow beam, a means of producing very short pulses (in order to give high discrimi-

---

[25] "War History," p. 39.

nation in range), a PPI display to give a general picture of the surroundings out to 40,000 yards, and a display that would take a sector 5 degrees on each side of any chosen bearing, and present it in range. It was necessary to produce a narrow beam that would scan this 10 degrees with a high rate of repetition, and this turned out to be difficult. The Admiralty gave a very high priority to this equipment, and a number of the staff were assigned to the project, with instructions to keep in touch with work on the K-band that was in progress south of the border.

It was a difficult project and a complex one, especially because the use of these very short waves was entirely new, and the necessary tubes and circuit components had not all been developed. The components of the "microwave plumbing" were much more difficult to make; it is interesting that the first mention of the project in the minutes of the Co-ordinating and Management Committee, on December 18, 1943, refers to "the expected difficulties in producing K-band fittings to close tolerance."

At the next meeting on January 3, 1944, "It was agreed that Mr. [H. R.] Smyth would be responsible for the proposed K-band project." Within the next few months about twenty people had been drawn into it, either full- or part-time. F. R. Park later became project engineer under Smyth; R. E. Bell worked on the antenna problem, the solution of which provided the most distinctively Canadian part of the equipment. It must be understood that no attempt was made to work everything out *ab initio*; whatever K-band components could be bought in the United States were thankfully incorporated. Not everything required was obtainable, and a local-oscillator tube for the receiver was produced in the vacuum-tube laboratory recently established in the Branch.

In May 1944 the design of the scanning antenna had progressed to the point where an engineered model was feasible, and the construction of this was begun. In September the Admiralty felt that the antenna was a success and that enough experience with the K-band had been obtained to justify asking the National Research Council to produce twelve fall-of-shot radar equipments complete with full sets of spares, for use in large naval vessels. This decision was undoubtedly connected with the great success of the type 268 radar already discussed. They designated the project as type 931. Work began at once, and by February 1945 a rough model of the entire equipment was taken to Halifax for tests on shell splashes produced by the coast artillery of the Canadian Army. The Canadian Navy provided a frigate so that the accuracy of ranging on such a target could be observed. Not only was the ranging on shall splashes eminently satisfactory, but it was found that at ranges of up to 10,000 yards the course of the frigate could be observed directly to within about 10 degrees. At 8,000 and

10,000 yards it was possible to observe separate echoes from the frigate and a buoy when they were only 50 yards apart. This was an indication of the excellence of the rapid-scanning antenna developed in the Branch.

Three officers from the Admiralty Research Establishment came to assist in the tests, and remained to help to determine the final design. The sets were built in the Radio Branch as a matter of urgency, parts being supplied by R.E.L. and other outside firms. Although the Branch had done much experimental work on a K-band transmitter-receiver unit, referred to as an "RF head," it was decided to use an equivalent unit already in production in the United States, and a number of these were purchased by the Admiralty.

In early June 1945 a complete equipment was ready for "user trials," which were carried out to the entire satisfaction of the Admiralty from a site near Scarborough, Ontario. In the absence of heavy artillery, shell splashes were simulated by firing charges of high explosive below the surface of the lake. In August, F. R. Park and C. H. Miller went to England to install the first equipment in a ship, but the war in the Pacific had ended and the installation was delayed until March 1946.

7.9 Other Projects for the Navy

The powerful pulses emitted by a radar transmitter would naturally interfere with communications receivers in the vicinity, and this was particularly important in a ship. A solution to the problem had been found at H.M. Signal School in the United Kingdom, and in the summer of 1941 the Royal Canadian Navy asked the Radio Branch to make such a device. The principle of this was to turn off the communications receiver electronically during the very short period in which each pulse was being emitted. Three such equipments were constructed, but when one was installed in a corvette in the fall of 1941, it was found that improved shielding of the radio equipment was equally useful. On the other hand, this apparatus was successful when installed at a ground station by the R.C.A.F.

A project in which the Branch was marginally involved was known as the panoramic receiver, a term that did not, as one might think, mean a receiver that looked all round the horizon. It was a receiver that scanned the frequency spectrum mechanically between 200 and 700 MHz. This was done by a motor-driven tuning capacitor. When a signal was detected it rang a bell. The reason for this development was a suspicion that enemy submarines might be using radar.

This receiver was given to a New York firm for development, but they were not told about its purpose. After sea trials of the second of two models about twenty were ordered from the Canadian Marconi Company through R.E.L. The main contribution of the Radio Branch to this project was the design of a suitable wide-band antenna for this range of frequencies, taking into account severe limitations of size and weight.

Finally there was a so-called "trainer" constructed for the Royal Canadian Navy in 1942 for use at the radar school at St. Hyacinthe, Quebec. This was needed for the purpose of demonstrating radar circuits to operators and maintenance men in training. "Circuits," says the "War History,"

> were built up on an open rack so that all tube base connections were easily accessible for instruction. All important points in each circuit were brought out to numbered terminals so that wave forms could be studied. The various circuits incorporated in the instrument demonstrated the functions of sweep circuits, ranger circuits, delay circuits and calibration-pip-forming circuits.[26]

These circuits formed the display on the cathode-ray tube of the radar apparatus, and must have been difficult for many of the trainees to understand without help of this kind.

While it was not constructed specially for the Navy, this is a convenient place to mention the apparatus that was developed in 1943 to facilitate the measurement of the beam patterns of radar antennas. At first these patterns were measured by having a technician walk back and forward across the beam, carrying a field-strength meter. Later on, under the direction of J. W. Bell and later, G. A. Miller, an automatic recorder was designed for this purpose, which made it possible to measure such patterns more quickly and more accurately.

---

[26] Ibid., p. 41.

Chapter 8

PROJECTS FOR THE ARMY

Following the plan of the previous chapter, the developments made at the request of the Army will be dealt with in the approximate order of their inception.

8.1 Coast-Defence Radar (CD)

The radar equipment for coast defence was the most entirely Canadian in concept of all the various types of radar developed during the war. In this section we shall consider only the earlier CD radar that operated at a frequency of 200 MHz; the later 3000 MHz (10 cm.) coast-defence radar will be dealt with in section 8.4.

Work on CD began in September 1939. It was hoped that the accuracy of such a radar equipment might be as great as that of an optical range-finder, say plus or minus 5 minutes of arc in azimuth and plus or minus 1 per cent in range. At first a wavelength of 60 cm. was tried, because a Western Electric Company radio altimeter that operated on that wavelength was available, and it was thought that this could be rapidly modified for the purpose.[1] Much time was spent during the winter of 1939-40 on antenna systems for such an apparatus, but because of the low power of the transmitter, and difficulties in a receiver for such a wavelength, it had become clear by June 1940 that a frequency of 200 MHz, that is to say, a wavelength of 1.5 m., would be preferable. Nevertheless, the time spent on the design of antennas was not entirely wasted, as it provided experience that was of great value in later developments. Four men from the Royal Canadian Corps of Signals were attached to the N.R.C. to help in this work.

---

[1] At the same time, experiments were begun on an air-to-surface-vessel radar (ASV) using the same altimeter. See p. 96 below.

A small number of the staff, who formed the nucleus of the later antenna group, began in August 1940 to develop the antenna for CD. A large rotating array was built at the field station, the mechanical design being done by H. E. Parsons, who had joined the staff in August and later became head of the mechanical engineering group. This construction was complete and ready for the electrical work by the end of October 1940. "This meant," wrote J. T. Henderson, "that the electrical measurements were made, and the aerial system evolved during the winter, and much credit is due to the workers of this group for their achievement under the most adverse working conditions."[2] Two of these men can be identified: J. W. Bell and G. A. Miller.

At the same time, other groups had been developing a transmitter and receiver for this radar equipment, and a preliminary demonstration was given for the benefit of staff officers on October 31, 1940. In the following February a second demonstration was given, and shortly after this a contractor began the construction of a seventy-foot wooden tower at Duncan Cove, Nova Scotia, near Halifax. Later a second installation was made at Osborne Head. With the agreement of the Army the Duncan Cove equipment was operated by the Navy.

About the middle of May 1941 a fitting-out party from the Branch arrived at Halifax to install the equipment. The names that come to light in the file[3] are those of K. A. MacKinnon, C. J. Bridgland, and A. Freeman, and the contents of the file illustrate the many dificulties and frustrations that the party experienced. In spite of this "by August the equipment was operating on a fairly routine basis,"[4] but tests and adjustments continued until in December 1941 and January 1942 accurate comparisons were made with the results of optical triangulation on target ships, with highly satisfactory results, indeed better than might have been expected from a radar equipment operating at 200 MHz. K. A. Mackinnon represented the Radio Branch at these tests.

However, in August and September 1941 the Radio Branch and the Navy conducted trials at Duncan Cove in order to compare three types of equipment: the new CD, the 200 MHz CSC radar,[5] and an air-interception radar just being developed in the United States by the Massachusetts Institute of Technology, which lent an apparatus.[6] This operated at 3000 MHz (10 cm.). Although this had only one-fifth the power of the CD radio, and although its antenna was only about

---

[2] John T. Henderson, "Progress Report for Period June 1939 to 1 January 1942," p. 14.

[3] Public Archives of Canada, N.R.C. file 45.2.55.

[4] "War History," p. 63.

[5] See p. 48 above.

[6] See p. 98 below.

FIGURE 8.1
Tower and antenna of CD radar

half as far above sea level, the results were comparable. This clearly indicated the potential advantages of the higher frequency for coast-defence work, but as the technique of using it was in an early stage, development of the CD was continued, especially that of a mechanical "displacement corrector" to take automatic account of the fact that the coast-artillery battery served by the radar was at some distance from it. The report (*PRA*-54, "Osborne Head CD Displacement Corrector") on this was signed by H. H. Rugg, R. D. Harrison, and G. R. Mounce, although J. T. Henderson[7] gives Harrison credit for the mechanical details.

Like other radar equipments, the CD set consisted (apart from the displacement corrector) of a transmitter, a receiver, and a rotating antenna. The antenna, a very large system, shown on its seventy-foot tower in Figure 8.1, was the distinctive and highly successful element of the apparatus. There were really two antennas, one for the transmitter and one for the receiver, each consisting of thirty dipoles and both mounted on the same rotating structure. An electrical device called lobe switching was used with the receiving antenna. This presented the azimuth operator with two overlapping "pips" on the cathode-ray tube, so that he might set on the minimum between them. Another operator determined the range, and both range and bearing (corrected by the displacement corrector) were transmitted to the battery by electrical devices called selsyns.

The rotating gear for this large antenna presented a mechanical problem. The success of the solution can be deduced from the fact that the measured probable errors were plus or minus 4.8 minutes of arc for the Duncan Cove set, and plus or minus 4.5 minutes of arc for the somewhat more powerful and slightly improved set at Osborne Head. The accuracy in range was plus or minus 29 yards and plus or minus 23 yards respectively, showing that all the other parts of the apparatus were equally satisfactory.

Even though it was clear by the end of 1941 that higher frequencies might give even better results, such sets for coast-defence purposes did not become widely available until 1944, and the CD sets continued in operation until 1945.

8.2 Anti-Aircraft Early-Warning and Gun-Laying Radar (GL IIIC)

The elaborate equipment known as GL IIIC was by far the most impressive of all the projects undertaken by the Radio Branch during the war. As finally constructed, this radar set consisted of two main parts, each mounted in a trailer with its own towing vehicle. The first, known as the Accurate Position Finder (APF) was the gun-laying and

---

[7] "Progress Report," p. 16.

range-finding part of the equipment, and operated at a frequency of 3000 MHz. It was housed on a trailer in a cabin that could be rotated on a vertical axis, and used separate antennas for transmitting and receiving, mounted at one end of the trailer so that they could be manoeuvred together in elevation, the whole cabin being rotated to determine the bearing of the signal. The external appearance of this part is shown in Figure 8.2.

The other part of this apparatus was known as the Zone Position Indicator (ZPI). The antenna for this, emitting elliptically polarized radiation at a frequency of about 150 MHz, rotated above a fixed cabin (Figure 8.3). The function of this was to show on a plan position indicator[8] the position of all aircraft within a range of 60,000 yards, so that a suitable target might be selected to be followed by the APF. The indicating portion of the ZPI was mounted in the rotating cabin,[9] the information being carried by connecting cables. One of the towing trucks carried a diesel power supply, and a third truck carried cables and spares.

This impressive convoy was, of course, the end-point of a great deal of experiment and designing carried on both at the N.R.C. and at R.E.L. with considerable urgency. It is no longer possible (and indeed it would demand too much space) to follow all the steps that led to the final production model, but from the "Progress Report," the "War History," and the minutes of the Co-ordinating and Management Committee the main events in the history of this development can be described in sufficient detail for our purpose. The official history of Research Enterprises Ltd.[10] affords an interesting view from the standpoint of the manufacturer.

The project began in a discussion between the Radio Section and the Canadian Army, in the persons of J. T. Henderson and Colonel H. E. Taber, in May 1940. The Army had learned that a large number of gun-laying sets would be needed if an acceptable apparatus could be made in Canada. On August 26 C. J. Mackenzie copied into his diary a telegram from Canadian Military Headquarters in the United Kingdom to National Defence Headquarters in Ottawa, which read in part:

> No great advantage to commence production in Canada present type i.e. GL Mark II and CD (150cm) . . . most desirable immediate steps be taken to provide manufacture in Canada of GL and possible CD equipments which are now under design and which should at not very distant date supersede or supplement existing types.

---

[8] See p. 5.

[9] This decision was severely criticized by both Canadian and British officers in England, for operational reasons. See p. 81 below.

[10] In J. de N. Kennedy, *History of the Department of Munitions and Supply, Canada, in the Second World War*, 2 vols. (Ottawa: King's Printer, 1950), vol. 1, pp. 407-41.

FIGURE 8.2
APF trailer

FIGURE 8.3
ZPI trailer

This telegram is interesting in two ways, first because it foreshadows the revelations that would be made by the Tizard Mission, and second because of the assumption that what would be done in Canada was simply to make equipment designed in the United Kingdom. This would in any event have been entirely impossible because of the great differences between electrical parts and vacuum tubes available here and across the Atlantic.

In spite of this telegram the Department of National Defence was requesting a 600 MHz set on October 7, and indeed nobody could have been sure at that time that the higher frequency equipment would be successful. In fact work on longer waves went on in parallel with that using microwaves until the following spring.

The development of a microwave GL radar by the Radio Branch had been decided at meetings held in Washington on September 16 and in Tuxedo Park, New York, on September 29 and 30. The last of these was the famous meeting at which the Tizard Mission demonstrated the cavity magnetron, and at which they made the British priorities clear to the Americans. The National Research Council, it was decided, should push ahead with the GL, while the Americans would take charge of the air-interception radar (AI)[11] and a long-range passive air-navigation device. These decisions had the full agreement of Mackenzie and Henderson.

Suddenly there was enough money in spite of the parsimony of the Treasury. On October 21, 1940, the acting President asked for $350,000 for radio research from the War Technical and Scientific Development Fund.[12] "Everything went through much easier than anticipated," his diary reads. "There were no objections at all." This was the day after the members of the British mission (except for Tizard, who had gone back to England) first came to Ottawa. Three days later, on October 23, they met officials of the Department of Munitions and Supply, and there was a good deal of discussion of the possible production of GL sets, with a clear recommendation that this should be of Mark III, the new microwave version. It was also suggested that special vacuum tubes might be made in Canada.

It was made evident to the visitors that they should go to Toronto and talk to Colonel Phillips, the President of R.E.L. This they did, and on October 28 Cockcroft, Bowen, and R. H. Fowler (the British Scientific Liaison Officer) returned to Ottawa full of enthusiasm and were entertained at dinner at the Chateau Laurier, with C. D. Howe and other ministers present, as Mackenzie noted in his diary. We must assume that they had come by the night train from Toronto; at any rate the day was spent in the Radio Section listening to Cockroft explain in

---

[11] See p. 98 below.
[12] Known also as the Santa Claus Fund (see p. 17 above).

detail the general characteristics of the proposed microwave GL set and the performance desired from it. This was written out and takes six pages, and it is remarkable how clearly Cockcroft had in his mind the general ways in which these requirements might be carried out. It is notable that this document begins by referring to the receipt of a cable from England, reporting that a range of 14,000 yards had been obtained on 10-cm. equipment.[13]

An example of the GL Mark I, which used longer waves, had been brought to Halifax from England, and next day Cockcroft, Fowler, and some U.S. and Canadian officers flew there to see it demonstrated. They also saw a demonstration of the Night watchman set already described in Chapter 7, which must have reassured the Englishmen as to the ability of the Canadian engineers.

At this time the Microwave Committee in the United States were also proposing to develop a GL radar. Fowler, writing to Mackenzie on November 4 from the British Technical Mission in Washington, urged that the most important contribution that Canada could make to this would be "for Henderson's group to push ahead on the highest possible priority with layout drawings for GL-3 for which ... they now have all necessary information."[14] He also hoped that Mackenzie would attend a meeting of the Microwave Committee in Washington on November 15. He did, and from this time onwards co-operation was cordial, if not entirely without friction.

By the end of November 1940 it had been decided to use a rotating cabin for GL IIIC. This apparently awkward arrangement was adopted because of the fact that nobody knew at this time how to design a waveguide with a low-loss rotating joint, so as to make it possible to rotate only the antennas. This alternative solution made it necessary to construct a very complex system of slip rings to convey power to the unit and to bring signals from the ZPI, and to convey target information to the gun batteries.

By this time R.E.L. had hired several mechanical engineers, and they gradually assumed responsibility for the mechanical design of the trailers. In February 1941 the N.R.C. placed contracts for three rotating trailers and three fixed ones. Meanwhile the electrical part of the equipment was under development in the Radio Section. It seems that the moving spirit in this was J. W. Bell, an ingenious and energetic man who had accompanied the members of the Tizard Mission on much of their inspection of plants in the United States.

The supply of magnetrons, cathode-ray tubes, and other vacuum tubes was a matter of serious concern. R.E.L. sub-contracted the

---

[13] The entire document, from Mackenzie, "War Diary," October 28, 1940, is reproduced as Appendix B.

[14] N.R.C. file S25-1-51, November 4, 1940. This interesting letter also deals with several other ways of co-operating with the United States.

making of magnetrons to the Northern Electric Company, who were able to supply satisfactory samples early in February 1941, "and considering that no one in Canada had any knowledge of 10-cm. magnetrons until September of 1940, this was a fine achievement."[15] Special cathode-ray tubes were being produced in Toronto by January 1942. For the 3000 MHz receivers, local oscillator tubes were being made there by April 1941, slightly modified copies of a British tube.

It was not certain until about the end of March 1941 that the experiments with 10-cm. radiation were going to be successful, but satisfactory echoes from planes at ranges up to 17,000 yards were obtained at that time, and the parallel work on 65 cm. was then discontinued. A group working at the University of Western Ontario under Professor R. C. Dearle and Dr. G. A. Woonton helped greatly by measuring radiation patterns of experimental antennas. Near the middle of May the final antenna system for the GL IIIC was decided upon—separate paraboloids for transmitting and receiving, mounted side by side and moving together in elevation.

Before the end of May the engineers were confident enough to promise a demonstration of the APF on June 17, and this date was met with the assembly of a complete working unit, although the method of presentation on the cathode-ray tube was improved in the months to follow. An early model of the ZPI was shown on the same day, but this had separate antennas for transmission and reception and was different from the final model in other ways.

On June 24, 1941, Mackenzie recorded in his diary that he was in Washington, trying to persuade the Commonwealth Scientific Office there of the importance of getting an order for R.E.L. to make 400 units of GL IIIC. He must have been persuasive, for on the twenty-seventh he noted that Colonel Phillips had phoned to tell him that an order for $26 million worth of GL IIIC radars was about to be signed. This covered 400 units that were to be bought by the United States for shipment to Great Britain on "lend-lease." Shortly afterwards the order was increased to 660 units. This shows a remarkable confidence in an equipment that was not at this time completely developed for production.

On July 23 a very successful demonstration was held at the field station before twenty-six American engineers and a number of U.S. Army officers. As Mackenzie told McNaughton on August 2, they "put on a demonstration and held an aeroplane accurately in the field for one and a half hours without losing it once, although it manoeuvred, changed elevation, distance, azimuth, etc., continually."[16] There is no doubt whatever that the Americans were vastly im-

---

[15] "War History," p. 45.
[16] Thistle (ed.), *The Mackenzie-McNaughton Wartime Letters*, p. 85.

pressed. The chief engineer of the Westinghouse Company said later that "his company . . . would not have believed that what we had done in nine months could have been done in under two years."[17]

This confidence in Canadian work was not shared to the same extent by the British. Professor Oliphant, who visited Ottawa in September, was very doubtful of the ability of Canadians to produce such a complicated apparatus, until he and Sir Lawrence Bragg themselves made tests on the prototype, after which he was converted, as Fowler told Mackenzie on October 20, adding in a postscript, "I would emphasize once more that all occasions for pessimism about the Canadian GL development are completely past,"[18] a statement that subsequent events proved to be unduly optimistic.

In all, five sets were constructed in the Radio Branch, except for items that could be contracted out. Besides the first model completed in June, one was sent to R.E.L. unassembled, another was sent to England in December 1941, one in 1942 to the U.S. Army, and a fifth kept in Ottawa as a standard. The set that was sent to England was loaded on flat cars under armed guard on a cold Saturday morning, December 6, 1941. It is very interesting to note that this was the day before the Japanese attack on Pearl Harbour that brought the United States into the war.[19] With it went five men from the Branch, led by F. H. Sanders. The others were J. E. Breeze, I. L. Newton, C. F. Pattenson, and A. K. Wickson. Breeze recalls[20] that before they went to England they had to enlist in the army and get transferred on loan to the National Research Council in order to circumvent the manpower regulations that were in force at the time. Then when this had been done, obtaining passports and being in the army at the same time led to administrative headaches. They left Halifax on December 10 in the steamer *Pasteur*, which had been built for service in the Mediterranean, not for the winter North Atlantic. The ship's manifest included, besides a large number of soldiers, "two trucks, two trailers, and five scientists." When they reached Greenock they found that no arrangements had been made for their further transport, and they stayed on board for five days and finally spent Christmas day on a train between Glasgow and London, with nothing to eat!

The GL IIIC was by far the most difficult problem that the Radio Branch and R.E.L. had to solve during the war. Apart from the complexity of the equipment, the fact that design and production had to proceed together made things very awkward from beginning to end. According to Kennedy, over 300 engineering changes were necessary

---

[17] Ibid.
[18] N.R.C. file B3.25.1.53.
[19] I am indebted to Mr. E. Webb for pointing this out.
[20] Interview by W. E. K. Middleton, November 14, 1978.

during the course of production.[21] The professionals in the Branch were mostly physicists, and those at R.E.L., fortunately, mainly engineers. In addition, there were the top-level clashes of personality already adumbrated in Chapter 6. The relative success of the programme was largely due to excellent liaison between the professional staffs of the two organizations, doubtless carefully watched over by Colonel Wallace. The "War History" tells how it was done:

> The Radio Branch endeavoured to make its designs conform to the general procurement situation which the Company had to deal with, and this close liaison was mutually beneficial. This was particularly true of the GL development, where great care was taken to ensure that the most up-to-date information was sent to R.E.L. At the beginning of August 1941 the Radio Branch began to send interim specifications on the construction of GL units to the Company . . . . These specifications were sufficiently complete to define the construction and performance of the unit.
>
> To keep still more closely in the picture, R.E.L. had a number of engineers working with us at N.R.C., and men were continually coming and going on specific phases of the program. The collaboration ensured that the model produced at N.R.C. was very close to a factory prototype, thus making for a minimum of delay in getting into production.[22]

Nevertheless there were those 300 engineering changes made necessary by the rapid progress of the art. Occasionally, as we may infer from the minutes of the Co-ordinating and Management Committee, a change that would have been an improvement had to be vetoed in order to prevent too great a disturbance of production.

Several accessory equipments were needed for the use of the GL IIIC, and the most important of these was undoubtedly the IFF (Identification, friend or foe). This is an apparatus that makes it possible to know whether the bright spot on the cathode-ray-tube screen is the echo from an enemy aircraft or "one of ours," in order that the anti-aircraft guns controlled by the APF may not fire on friendly planes. To accomplish this, the latter were fitted with "transponders" which swept through a band of frequencies from 157 to 187 MHz and returned a coded signal when the interrogating signal from the radar station was received. Frequencies of 165 and 171 MHz were assigned to GL IIIC. The transponder, installed on all allied aircraft, was not part of the responsibility of the Radio Branch.

It was at first thought that IFF would be added to the GL in England, and some development was done there, but later on it was requested that the GL IIIC be provided with IFF in Canada, and at a meeting held at the War Office in London on April 1, 1942, and

---

[21] Kennedy, *History*, vol. 1, p. 430.
[22] "War History," p. 49.

attended by F. H. Sanders and others from the Radio Branch, it was proposed that part of the development should be done at the N.R.C., namely, the antenna and the display. The interrogator and the responsor were to be made by R.E.L. from English designs.

This project went very slowly at first; it was not until August that prototypes and information began to arrive in Canada. It was decided that the IFF display would be added to the ZPI part of the GL IIIC, with a separate cathode ray tube for IFF. A separate directional antenna mounted on the APF trailer would be rotated by the operator to interrogate any aircraft that appeared in the PPI display of the ZPI. Later the British demanded that the IFF display should also appear on the APF.

By May 1943, as a result of acceptance tests, a specification had been drawn up. A prototype suitable for quantity production was made by R.E.L. and taken to England by people from the Radio Branch in December 1943. With orders for 655 sets (one for each GL IIIC), manufacture started in the summer of 1944 after further modifications, and shipments began in October, but the war was going so well that the British cancelled their part of the order. In all 350 sets were produced for Australia, South Africa, Russia, and Canada. Production was finished in September 1945.[23]

Another device requested by the Army was an apparatus to facilitate the training of people to operate the APF part of the GL IIIC. The usual way to teach this difficult task was to fly aircraft in the vicinity and have the operators practise ranging on them. At the Canadian Army Radar Training School, Debert, Nova Scotia, this procedure was found very unreliable, especially because of the east coast weather; it was also very expensive, but nobody was thinking about that in 1942.

The Army, therefore, asked the Radio Branch to develop a training device with the following characteristics: (1) it should use the APF trailer and provide for tracking in range, bearing, and elevation; (2) there should be no modification of the APF set; and (3) what the trainee would see on the cathode-ray tube should, if possible, be exactly what he would see when tracking an actual aircraft.

Experiments were begun in March 1942, and by July an apparatus was completed and ready for test. A carriage was arranged to move along a 200-foot length of track 35 feet above the ground. On the carriage was a receiver that picked up the pulse from the APF and a low-power oscillator that re-radiated it after a delay that could be adjusted from the ground. The antenna was a dipole in a 16-inch paraboloid, which produced circularly polarized radiation so that the carriage could be tracked in both bearing and elevation, while the adjustable delay provided for tracking in range. The signal could be

---

[23] Ibid., pp. 51-54.

faded at will in order to make the display more realistic. The errors made by the trainees were determined by the use of a telescope aligned with the antenna of the APF.

The slowness in getting this equipment into use can be better recounted than explained. The whole apparatus was packed for shipment in the fall of 1942 but was sent to Debert, Nova Scotia only in February 1943. By this time it had been decided to move the school to Barriefield, Ontario, but the equipment (with a track 385 feet long) was not installed there until January 1944, when four men from the Radio Branch spent a fortnight at Barriefield. Later an illuminated optical target was added so that the errors made by the operators could be observed by night as well as by day.

The same sort of apparatus, but mounted on a vertical track, was provided to R.E.L. in June 1943 for the alignment of the GL IIIC. This had been done by extensive tests on flying aircraft, so that the new device saved much time and expense, and also improved the accuracy of the radar set. The device was used with success at R.E.L. during the latter half of the production run.

It would be pleasant to report that the GL IIIC was satisfactory in every way, but unfortunately this was far from the truth.[24] The difficulties were partly technical and partly administrative. Let us deal with administrative matters first.

Even before the first set left Canada the accountants had their say. On October 27, 1941, Mackenzie wrote to McNaughton:

> The scientific officers in the various stations in England have been most anxious to get our set sent over at the earliest moment. As they only wanted it under loan, and the commercial cost is about $60,000 and the cost to us about $100,000 or $120,000, we decided that the best way would be to send it to you, and the financial accounting then could be done in Canada between ourselves and the Department of Defence here. Then you could permit it to be tested and examined by the various people in England who are particularly interested.[25]

It appears that Canadian headquarters in England did not in fact permit these "various people" to test the set, which "went to Petersham for 2 months operational trials and then to the Canadian Army for training, rather than to A.D.R.D.E.[26] for detailed technical study and constructive criticism."[27] It was in fact March 1943 before

---

[24] The British experience with the apparatus can be pieced together from three large files in the Public Record Office at Kew: AVIA 22/1473, AVIA 7/3262, and AVIA 7/3263. Dr. Timothy N. Metham was engaged to examine these and other files and produced a remarkably comprehensive report from which these observations are derived.

[25] *The Mackenzie-McNaughton Wartime Letters*, pp. 97-98.

[26] The Air Defence Research and Development Establishment at Malvern.

[27] J. D. Cockcroft in a report on file AVIA 12/189.

A.D.R.D.E. received a set, and by this time sets had gone to Anti-Aircraft Command and some of these were giving trouble.

The British were also having administrative difficulties of their own. The War Office had told A.D.R.D.E. that they should advise on the operational use of the equipment but that they should not try to modify it. The result of this is well described by Dr. Metham:

> Thus, to begin with, all problems had to be referred back to Canada for action. The time delay was clearly disastrous, and when matters came to a head, ADRDE did begin to advise on, investigate, and detail instructions for modifications, although they still regarded these modifications as "temporary" pending Canadian action. At the same time, A.A. Command (Anti-Aircraft Command), seem to have taken things into their own hands and prepared their own modifications. Once they had made the decision to go it alone, A.A. Command, in fact, seems to have been quite capable. However, this was clearly not the way things were done in the British Army and an acrimonious correspondence developed over the question of "authority" for modifying equipment.

This matter of modifications leads us to ask why they were considered necessary, and so to a discussion of the technical shortcomings of the equipment. It should be stated at once that only a few of these related to design, and that most of them were the result of inadequate quality control at R.E.L.,[28] failure of commercial components, an insufficient provision of spares, and inadequate training for the operators. Nevertheless there was one general feature of the design that was found to be unsatisfactory in the field. This was that the two components, ZPI and APF, had to be used together. This caused difficulty because of the absence of suitable Canadian IFF attachments, so that British IFF sets had to be used, and these interfered with the ZPI. A change in the frequency of the ZPI to 145 MHz was proposed, but the modification to the antenna took a long time and the British GL Mark II was gradually substituted for the ZPI.

A matter of particular concern to operational officers was the location of the PPI display in the APF trailer. On September 5, 1942, Lieutenant-Colonel F. F. Fulton of Canadian Military Headquarters, London, wrote to Ottawa[29] that after discussion with several officers he could see no single reason for this; instead, he gave seven reasons for its removal to the ZPI. All of these seem cogent to the present writer. In transmitting this letter, the Liaison Officer, Professor Shenstone, agreed with them but said that the design had evolved to meet requirements suggested by the British. It can only be supposed that these suggestions had been misinterpreted.

The question of quality control at R.E.L. is illuminated by the reports of Major P. W. Larsen of the Inspection Board. For example, on

---

[28] See p. 44 above.
[29] AVIA 7/3263.

October 5, 1942, he stated that of fourteen sets submitted in September, nine were rejected.[30] Some of the commercial components gave trouble, for example, small transformers and blower motors. If the blower that cooled the magnetron ceased to function properly, the magnetron was destroyed, putting the set out of action.

There were undoubtedly some design faults in the GL IIIC that were due to insufficient appreciation of the stresses that would occur in the field. The question of possible electrical and mechanical injury to the operators does not seem to have been given enough consideration. Training and maintenance manuals prepared in Canada seem to have been incomprehensible in the United Kingdom. M. J. Neale of the Radio Branch was in England for much of 1943 to help sort out these problems, and he was joined by W. H. Happe in July for a few weeks. Hackbusch and two other men from R.E.L. were also there at this time. The quality of the sets coming from Canada gradually improved, but the question of the ZPI remained. As for the earlier sets, improvements were severely hampered by jealousy between various British departments. The operators of the equipment usually did not seem to recognize that it was a precision apparatus which demanded careful adjustment and maintenance. When these conditions were assured the GL IIIC gave satisfactory service.

Nevertheless it was not to be used in the recovery of Europe. On September 9, 1943, Canadian Military Headquarters in London noted[31] that the Canadian Army overseas would use the British GL III and not the Canadian, for four reasons: (1) it was War Office policy to use only the British equipment in the field; (2) several British heavy anti-aircraft regiments would be with the Canadian Army; (3) base ordnance depots would be under British control; and (4) there would be difficulty in operating and maintaining two different types of equipment used in the same organization for the same purpose. It must be admitted that the last three of these reasons seem entirely valid.

8.3 Microwave Zone Position Indicator (MZPI)

The ZPI described in the previous section was susceptible to enemy jamming, and it was very difficult to use in mountainous country because of interference from "permanent echoes" in such terrain. Discussions between British and Canadian Army officers led on April 9, 1943 to an unofficial request to the N.R.C. to design an equipment that would not have these defects. Immediate action was taken but at low priority until an official request and performance specifica-

---

[30] Ibid.
[31] AVIA 22/1473.

FIGURE 8.4
MZPI trailer

tions were received from the Canadian Army in December 1943. It was desired to have a PPI display with optional ranges of 40,000, 80,000, and 120,000 yards, with a bearing accuracy of plus or minus no more than 2 degrees, and coverage from the horizon to 70 degrees in elevation. The radar was to use the S-band.

An experimental equipment was tested at Ottawa in January 1944, and after a great deal of work early in 1944 a modified equipment was sent to Vancouver in order to test it in mountainous country. The results were so satisfactory that the project was given top priority; however, as a result of revised British specifications it was December 1944 before the N.R.C. prototype could be shipped to England for trials, which were eminently successful.

The equipment as finally produced is shown in Figure 8.4. We shall not go into the details of this elaborate and extremely sophisticated apparatus, but the peculiar antenna shown in the photograph should receive some comment. This was based on an original design by W. H. Watson of McGill University. There were really two antennas, one consisting of six slotted waveguides stacked vertically and the other having three waveguides tilted about 30 degrees from the vertical. By appropriate switching of the feed this gave coverage up to 70 degrees in elevation. The whole assembly could be folded down for transport.

A great deal of thought was given to the problem of making this equipment suitable for use in both tropical and arctic climates, both as regards the functioning of the apparatus and reasonable comfort for the operators, matters which had been given insufficient consideration in most wartime equipment.

The MZPI radar was developed during the last phase of the war. It was eminently successful except that a greater range was later considered desirable by the Canadian Army, and a redesign to increase its range was in progress after the period covered by this book.

8.4 GL IIIC Star

By the end of 1942 "shortcomings of the GL IIIC were already evident,"[32] and in January 1943 it was decided to design an improved version. At this time there was a possibility of an order of 100 sets for the United Kingdom and 10 for Canada. The modification of both the ZPI and APF sections of GL IIIC was begun at a fairly high priority in February by the Scientific Research and Development Section, the Mechanical Engineering Section, and the Army Section. The project eventually came under the general supervision of J. E. Breeze. By the autumn of 1943 the design had reached an advanced stage and a final

---

[32] "War History," p. 59.

specification had been drawn up. However, the progress of the air war in Europe led the British to withdraw their support for the project, which would have been discontinued entirely except for the sustained interest of the Canadian Army. In the event it was continued at a very low priority, but when in December 1944 the Branch asked the Army to countermand it, they found "that the Canadian Army [did] not wish to cancel the GL IIIC programme, either at the N.R.C. or at R.E.L."[33] In spite of this, work was stopped at the end of the war before actual flight trials had been performed. It is clear that in 1944 the Radio Branch estimated the priority of this equipment more accurately than did the Canadian Army.

As the most visible modification was the substitution of a single rotating paraboloid for the antennas and the rotating cabin of the GL IIIC, we may suppose that the desire of the Branch to end the project was due to their study of a superior equipment, the well-known SCR-584, that had come into production by this time in the United States.[34]

8.5 Searchlight-Control Radar (SLC)

The possibility of using radar to control the pointing of a searchlight must have been evident at an early stage in the development of the technique. The idea that a searchlight might be pointed at an enemy and only then turned on had obvious attractions to those engaged in defence against aerial bombardment. Early in the war the British succeeded in improvising a searchlight-control radar by using components of their ASV[35] radar.

In March 1941, after a meeting with Army officers in February, the Radio Branch began experiments with a view to adapting the "Night watchman"[36] radar to searchlight control. There were two such equipments, one operating at a wavelength of 140 cm., the other at 85 cm. The former proved the better, and as the CSC[37] radar was then coming into production at R.E.L. it was decided to use the transmitter and receiver of the CSC for the SLC equipment. The Army placed an order for fifty units, but because of the very great pressure to get the GL IIIC into production very little was done during 1941.

In January 1942, in spite of information that the utility of radar control of searchlights was a matter of debate in the United Kingdom, the Branch was asked by the Army to develop a new lightweight SLC radar operating in the microwave region, with the capability of both

---

[33] C. and M. Committee Minutes, December 30, 1944.
[34] Mr. J. E. Breeze informs the writer that this supposition is correct.
[35] See Chapter 9.
[36] See p. 47.
[37] See p. 48.

searching for and tracking aircraft. Such a unit was in fact built, and is shown in Figure 8.5, the scale of which can be inferred from the information that the paraboloidal antenna was 48 inches in diameter. The cabin, which had to accommodate the observer as well as some of the electronic components, would seem to have been somewhat cramped.

The "War History" devotes a good deal of space to a narrative of the attempts made during more than two years to make an apparatus of this kind work satisfactorily. This story will not be repeated here, especially as it was decided in March 1944 to abandon the project rather than embark on the complete redesign that experiments had shown to be essential. In point of fact the project had never had a very high priority. It is interesting to note that one of the reasons that a complete redesign would have been necessary was that when the cabin was given a sufficient angular velocity most operators suffered from motion-sickness.

8.6 Coast-Defence Gun-Laying Radar (CDX)

In the discussion of the coast-defence radar CD it was noted that experiments carried out in August 1941 led to the conclusion that higher frequencies would give even better results than the very successful CD equipment that operated at 200 MHz. In January 1942 the Canadian Army asked the Radio Branch to develop a coast-defence radar with a high degree of discrimination between targets, and it was clear that this should use a wavelength of 10 cm. On April 9, 1942, in an astonishing demonstration of confidence in the Radio Branch, the Army ordered forty-one such units from R.E.L., less than a fortnight after a design conference had been held at the N.R.C. to discuss the specifications for the new equipment, to be known as CDX. On May 12 the actual performance specifications were received from the Army, dealing with such things as maximum and minimum ranges, allowable error, rates of change of bearing, and minimum rates of tracking. The apparatus had to include a "displacement convertor . . . capable of providing continuously and automatically fire control data corrected for the displacement between CDX and gun site."[38] It was decided to make use of whatever components of the GL IIIC could be incorporated without alteration.

A preliminary model—what the engineers call a "lash-up"—was made without the displacement converter and was tested at Osborne Head, Nova Scotia between October 15 and 20, 1942. Although the CDX was mounted only about half as far above sea level as the CD, the range and accuracy were better. "The absence of clutter and spurious

---

[38] "War History," p. 67. The spelling *convertor* is a solecism.

FIGURE 8.5
Searchlight-control radar

echoes so prevalent on 150-centimetre equipment was one of the most encouraging proofs of the superiority of the S-band unit."[39] The results of the tests led to great activity in the design and construction of an engineered apparatus.

The general appearance of the CDX radar is shown in its experimental form in Figure 8.6. There were two 4-foot paraboloid antennas, one for transmitting and the other for receiving, mounted on a housing that contained the transmitter and receiver. The specifications called for all this to be rotatable through an angle of 300 degees only, so that slip rings were not required. In the production model (Figure 8.7), the rotating equipment was mounted on a low pedestal on the roof of the operating room, which in an actual installation would be of reinforced concrete.

The prototype set was finished in February 1943 and tested in Ottawa in Februry and early March and at a coast artillery battery near Halifax in April. The results of the tests were eminently satisfactory and demonstrated the superiority of the microwave equipment. It was also very much more compact, as will be seen from the David-and-Goliath picture reproduced as Figure 8.8, perhaps symbolic of the continual "miniaturization" of electronic equipment that has gone on ever since.

After these successful trials, design specifications were completed at the N.R.C. in May. Meanwhile, the prototype was shipped to R.E.L. and set up at their field station at Scarborough on the shore of Lake Ontario. Apart from liaison with R.E.L., no further work was done on CDX at the Radio Branch. The favourable course of the war caused the original order to be reduced from forty-one to fourteen, but five additional sets were built for the Russians.

Rather surprisingly, work on IFF equipment for the CDX radar was not started until March 1944; it continued by way of liaison with R.E.L. up to the end of 1945. A somewhat optimistic maximum range had been proposed. Trials at Scarborough in November 1944 showed that in order to be able to interrogate vessels to the visual horizon there would have to be about ten times the power and a large billboard antenna. There is no record of such a development.

Another complicated addition to CDX was a device to present automatically to the battery commander the positions of shell splashes relative to that of the target. The original design was suggested by the Directorate of Artillery of the Canadian Army in February 1945, and a prototype built at the N.R.C. was tested at Scarborough, Ontario in June. Various improvements resulted from the tests. The final model was tested at Esquimault, British Columbia in

---

[39] Ibid., p. 68. "Clutter" is a term for signals on the cathode-ray tube resulting from reflections by waves in the sea, etc.

FIGURE 8.6
CDX radar, experimental form

FIGURE 8.7
CDX radar, production model

FIGURE 8.8
The two coast-defence radars

October 1945 and performed so well that the Army itself began in January 1946 to build one for each of its CDX radars.

The project shows the extent to which co-operation between the Radio Branch and the Canadian Army had developed by this time. The details of the equipment involve a number of highly technical concepts and will not be given here; it must suffice to quote from the "War History" to show what it did: "in addition to providing an instantaneous, pictorial presentation of splash errors to the Battery Commander the unit provided him with a ready check at all times on the accuracy with which the CDX bearing operator was following the target."[40]

One of the features of the CDX radar was a means of indicating the rates of change of the range and bearing of a target that was being followed. It was obvious to the Army that there would be a great advantage in having a computer built into the apparatus to convert these data automatically and continuously into information about the course and speed of the target. In May 1944 several mechanical solutions to the problem were tried by the Army, but it soon became clear that an electrical device was essential. Unfortunately the project was given a low priority and very little work was done on it.

8.7 Forward-Area Warning Radar (FAW)

At the beginning of 1942 the Canadian Army asked that a radar similar to the ZPI should be provided as a separate unit, so that it could be installed independently of the GL IIIC to provide early warning of the approach of aircraft to troops in forward areas. Another use for such equipment was as a supplement to the GL at sites where the ZPI was not fully effective because of the lay of the land. The separate ZPI would then be sited at a more favourable place in the vicinity.[41]

It was easy to produce a ZPI with its own gasoline-driven generator, installed in a truck and its trailer. During February and March 1942 such an equipment was assembled from components then being produced at R.E.L., tested, and sent to England for further evaluation. An operational specification was then prepared for an FAW set, which was expected to have a range of 120,000 yards on aircraft, with an accuracy of plus or minus 1,000 yards, and an accu- racy of plus or minus 2 degrees in bearing. There were to be no gaps in its field of view, and it had to be mobile and capable of being put into operation in fifteen minutes.

---

[40] Ibid., p. 74.
[41] Note that the desirability of having the ZPI as a separate unit was soon to be emphasized by the British. See p. 81 above.

The design of such an apparatus began in August 1942, and a prototype was given flight trials in December and turned over to R.E.L., who completed a production prototype in August 1943. Flight trials at Toronto and on the west coast showed that while the set was an improvement on the original ZPI, it was not very much better, and by that time another kind of early-warning radar[42] was being developed, which promised to be superior; the project was stopped after two production prototypes had been made.

8.8 Counter-Bombardment Radar (CB)

A mortar is a relatively compact and easily concealed piece of artillery that sends up bombs on a high trajectory. Mortar fire may produce many casualties. The obvious defence against it is to destroy the mortars by artillery fire or aerial bombs, and to do this they must be accurately located. It naturally occurred to the army that radar might be able to plot the trajectory of the mortar bombs so that their point of origin might be determined.

Early in 1944, experiments for this purpose were under way in England using 10 cm. (S-band) radiation. In Canada, trials using 3 cm. (X-band) waves were undertaken in April and May 1944 with unsatisfactory results. It was evident that a shorter wavelength should be tried. It became possible to borrow a K-band (1.25 cm.) radar from M.I.T., although this was a laboratory model with low power output. Trials at Wainwright, Alberta, at Meaford, Ontario, and at Petawawa Military Camp gave promising results. At Wainwright echoes from swarms of small gnat-like insects were observed; this was probably the first radar observation of insects.

Early in 1945 the type 931 radar[43] was being tested at Halifax, and the opportunity offered itself to see whether it could track mortar bombs. The results were so good that in February 1945 the Radio Branch began to design a special radar apparatus to be mounted on a full-track vehicle. The principle of the design was as follows: a narrow beam would make a horizontal scan at two elevations above the horizon in rapid alternation. In this way two points on the trajectory of a rising mortar bomb could be established, and extrapolation down to the ground would give the position of the mortar with sufficient accuracy. The engineering problems of producing a lightweight set for this purpose were formidable, and were carried on throughout 1945, with somewhat reduced priority after the end of the war. The project was continued into peacetime and was still active after the period covered by this book.

---

[42] See p. 101.
[43] See p. 63.

8.9 Range-Only Radar for Bofors Predictor

I shall refer briefly to a project that was evidently given such a low priority that it never even reached the stage of an experimental model. This was to be a microwave (3 cm.) radar indicating only the slant range of a target aircraft, to be mounted on the top of the predictor used with Bofors guns, so that its parabolic antenna would constantly point at the target as it was being followed by the operator of the predictor. The range shown by the radar was to be fed directly to the height mechanism of the predictor.

In the spring of 1943 the development of such an equipment was started in England.[44] For some reason the Canadian Army, knowing about this, asked the Radio Branch to develop a comparable apparatus, but the work proceeded at such a low priority that it was not until February 1944 that any of the engineers of the Branch took time to see a Bofors shoot at an army school, a visit that "proved a great aid in actually visualizing the problem."[45] In March, work on the project was suspended, for by this time the development was nearing completion in England.

8.10 Interference-Measuring Set

An interesting example of the way in which the Branch could solve small but annoying problems is furnished by an instrument to measure the radio interference caused by the dynamotors used to supply plate voltages in radio-communications equipment made in large quantities for the Army. This problem was put to the Branch in August 1943 by the Inspection Board of the United Kingdom and Canada. Dynamotors with commutation problems caused excessive noise, known as "hash," in the receivers to which they were connected, and as these devices were being made in very large numbers, a quick "pass-fail" production test was desired.

The Branch decided very quickly that not enough data were available for the solution of the problem. In particular, nothing was known about the interference spectrum of these dynamotors. Measurements of this were made for the Branch by the Radio Division of the Department of Transport; it was found that there was a peak in the spectrum at a frequency of 5 MHz. After this it was fairly simple to provide a stable receiver tuned to this frequency and a device attached to the output of this receiver that turned on a green light for an acceptable dynamotor and a red one if the machine had to be rejected. The

---

[44] On August 16, 1941, it was suggested at a meeting of heads of groups that a range-only radar should be developed for this purpose. Nothing seems to have been done at that time.

[45] "War History," p. 84.

instrument was used in the factory to inspect all subsequent production.

Chapter 9

PROJECTS FOR THE AIR FORCE

In this chapter we shall deal with a number of projects that the Radio Branch undertook at the request of the Royal Canadian Air Force. We shall also include one project, useful to a much wider circle, which progressed throughout the period—the provision of standard radio and audio frequencies.

9.1 Airborne Radar, Air-to-Surface Vessel (ASV)

The very first problem proposed by the Services to the Radio Branch was the development for the R.C.A.F. of an airborne radar for the detection of ships or surfaced submarines. This project was officially endorsed in September 1939. It so happened that the Western Electric Company in the United States was manufacturing a type of radio altimeter using frequency-modulated radiation at a wavelength of 67 cm. The Radio Section thought that they could save time by modifying one of these altimeters to emit pulses of waves instead of continuous frequency-modulated radiation, and by adding a cathode-ray-tube indicator.

It was November before the altimeter arrived, and by the end of that month echoes had been observed from aircraft and buildings at short ranges. The work on the coast-defence radar (CD)[1] was going on at the same time, and only in June 1940 was it possible to make any flight trials. In the interval, however, another transmitter and receiver had been built, of about the same power but operating on a wavelength of 1.5 m., with which there was much more experience in receiver design. It was recognized that the antenna system could be made much more compact if the shorter wavelength was used, and

---

[1] See p. 67 above.

the experiments on 1.5 m. were intended to determine which of the two wavelengths should be adopted. These experiments were useful in the design of the CD radar and of other sets working at frequencies below 200 MHz.

Because of the many demands on the time of the staff, it was decided to try to obtain British 1.5-metre equipment for the R.C.A.F., or at least to procure a sample from which a Canadian design could be made. Unfortunately, there was a very long delay in its arrival. It came to Ottawa after the Tizard Mission had arrived, but the Branch had it for only forty-eight hours, after which it was sent to Washington to be fitted in a United States aircraft as a demonstration. It may be imagined that the staff worked very hard to get as much information as possible during those two days. This sample never did come back to Ottawa, and in spite of strenuous efforts by the Liaison Officer in London, it was not until January 19, 1941, as a result of a personal intervention by C. D. Howe on a visit to London, that prototypes of ASV, AI, and IFF arrived at the Laboratories.

When it was known that they were on their way a group of engineers and technicians was hired by R.E.L. and sent to Ottawa to make a copy of the British ASV Mark II. This had to be physically and electrically interchangeable with the British set, but using Canadian and American tubes and components. With the full co-operation of the Branch a satisfactory model was constructed by the end of February 1941. After this the R.E.L. group were gradually transferred back to Toronto, some of them remaining in Ottawa until the factory at Leaside was ready. During the course of the war R.E.L. produced several thousand of these sets for use by the Allies. "The larger part of this production was sold directly to the U.S.A. for installation in American planes, thus filling a gap in the American production program."[2]

It will be seen that the Radio Branch did not play a major part in the design of this equipment. However, it did a good deal of work on the design of antennas for ASV, and for ground beacons installed by the R.C.A.F. to work with ASV.

9.2 Ground Radar for Detection of Aircraft and for Control of Interception (CHL/GCI)

The CHL/GCI was another equipment that has a small place in this history only because the facilities of the Radio Branch were used by a group of engineers from R.E.L. in the winter of 1940-41 before the factory at Leaside was completed. These engineers were able to make use of the experience already gained by the Branch on other high-

---

[2] "War History," p. 91.

powered radar apparatus. As in ASV, the equipment was essentially a redesign of British equipment using North American components.

9.3 Airborne Radar for Aircraft Interception (AI)

We have already noted[3] that it was decided at meetings in the United States in September 1940 that a microwave AI system, for which the British had an urgent need, should be developed by the Americans. It was also decided that the Radio Branch should co-operate in the project. On December 16 Henderson and two others visited M.I.T. and arranged for a group of six men from the Branch to be transferred on loan to assist in the AI project. This had been agreed to by the acting President of the N.R.C. The purpose was to ensure that the Branch kept in touch with developments in microwave technique.

The six men were at M.I.T. for practically all the first five months of 1941, witnessing and helping in the truly remarkable blossoming of microwave research in the United States. During this time an AI antenna was sent to Canada and installed by the R.C.A.F. in an aircraft that was later sent to England with a complete AI equipment included, for the purpose of comparing the British and the American apparatus. One engineer from the Branch went with the American party sent with it. A second AI set was sent to Ottawa during the summer and used in the comparative tests near Halifax mentioned in Chapter 8.[4] Apart from this the Branch had very little to do with this project.

9.4 Long-Range Early-Warning Radar (LREW, VEB)

Even before the outbreak of war, as soon as the British developments in radar were known in Canada, the Royal Canadian Air Force set forth a requirement for equipment designed to give the earliest possible warning of aircraft approaching at any height from any direction. After the fall of France there seemed to be a real possibility of airborne attacks on Canadian cities, and the matter was felt to be urgent. Unable to purchase such equipment in the United Kingdom, the R.C.A.F. turned to the National Research Council. Instead of merely copying the British CHL apparatus, it was decided early in 1941, in consultation with the Radio Branch, to ask for a powerful station with a PPI display, and also with the capability of determining the height of the aircraft. The station was to be able to "see" 90 per cent of all aircraft within a range of 100 miles, with a range accuracy of plus or minus a quarter of a mile and a bearing accuracy of plus or minus 1 degree. The height-finder had to be able to furnish angular eleva-

---

[3] See p. 23.
[4] See p. 68.

tions to plus or minus a quarter of a degree at all elevations between 2 degrees and 30 degrees. All this had to be done with only one tower and only one transmitter and receiver.

This was an extremely tall order, and eventually the accuracy requirements were relaxed to some extent, and as long as detection and height-finding could be done simultaneously, separate transmitters could be used for the two functions. The original limitation suggests that the Service officers who wrote the requirements knew little about the state of the art.

The equipment as finally installed at the field station south of Ottawa was very impressive indeed. Its 200-foot tower (Figure 9.1) was an object that must have caused much speculation to nontechnical inhabitants. As it turned out, this experimental installation was the only one of its kind, for by the time it was completed microwave techniques had progressed to the point where a microwave early-warning radar was superior.

The tower itself was not installed without trouble. Even in wartime it was necessary to obtain permission from the Civil Aviation authorities for an obstruction only a few miles from the Ottawa airport, and on May 23, 1941, we find Mackenzie writing to Air Vice-Marshal L. S. Breadner to ask for his support in obtaining the required authorization.[5] By September a contract had been let to the Ajax Engineering Company, but they seem not to have investigated the soil conditions at the site, for in November they found that they needed much more elaborate foundations and asked for more money. However, the tower was up before the end of January 1942, except for an elevator, which was not finished until June. This elevator, says the "War History," "has proved of inestimable value in the experimental work, but would obviously be a luxury in operational installations."[6] We have indeed seen how few luxuries there were at the field station.

On the top of the tower was mounted a large billboard antenna for the PPI display, rotating continuously at 4 r.p.m. This was built during the spring of 1942, mounted on a fifteen-foot tower, and thoroughly tested during the summer before being erected on the tall tower in October. Wisely, at the suggestion of J. T. Henderson, the Branch forbore to risk the lives of its personnel in this operation, and let a contract to professional riggers.

Near the bottom of the tower were two small buildings, one holding the powerful transmitter, the other the receiver and displays. The transmitter, developed by the Radio Branch, had four large transmitting tubes and supplied pulses with four or five megawatts peak power.

---

[5] N.R.C. file M6.30.14.
[6] "War History," p. 96.

FIGURE 9.1
Tower and LREW antenna

The PPI display was designed by H. E. Ferris and was something of a "breakthrough." The output of the receiver was applied to the intensity grid of a twelve-inch cathode-ray tube with a long-persistence screen. The original part of the idea was to rotate the sweep by rotating a deflecting magnetic field around the tube; this was done very neatly by a pair of position motors (selsyns). One rotated with the antenna and the stator coils of another were placed round the neck of the cathode-ray tube. "This was the first application of this PPI sweep system," says the "War History" with pride, "which was . . . eventually used in many radar sets of the United Nations."[7] The relatively long 8-microsecond pulses, the narrow bandwidth of the receiver, and the 12-degree beam of the antenna combined to give a display that was particularly effective at long ranges, for which it was primarily intended. We are told that aircraft were frequently picked up and tracked well beyond 200 miles.

In addition to the LREW antenna, another array was mounted on the south side of the tower (Figure 9.2). This provided a beam of variable elevation (VEB) for height-finding, 90 degrees wide horizontally but only 4 degrees vertically. The elevation of the beam was varied by sweeping the frequency of the transmitter between 75 and 95 MHz automatically every five seconds, at the same time varying the tuning of the receiver. In an operational equipment there would have been four such arrays, one on each side of the tower, to cover the entire horizon.

Before the VEB array was mounted on the tower it was set up horizontally to facilitate testing, for this position gave a beam that could be varied in its horizontal direction by sweeping the radio frequency. It was late spring of 1943 before the VEB was operating satisfactorily, and the whole LREW system was then demonstrated to the R.C.A.F.

Some work was done on a VEB system operating in the region of 400 MHz, but this was abandoned in favour of the microwave early-warning and height-finding systems that will now be considered.

9.5 Microwave Early-Warning Radar (MEW)

Before the war and up to about the end of 1942, frequencies of 200 MHz or less were everywhere employed in radar apparatus designed to give early warning of approaching aircraft. This was because the long ranges that were desired made it necessary to use very high power, and at the same time the design of receivers for the lower frequencies was a familiar technique. On the other hand, the radiation pattern of the antennas that had to be used was such as to make it

---

[7] Ibid., p. 97.

FIGURE 9.2
Antenna for VEB

possible for low-flying aircraft to approach unobserved, and the angular resolution was limited by the size of the antenna structures necessary at metre wavelengths.

The very great advantages of the PPI display for this application were beyond doubt, and the LREW radar described in the previous section provided such a display, but at the cost of a very large, heavy, and costly rotating antenna. By the middle of 1942, magnetrons capable of providing pulses of more than 300 kilowatts at a wavelength of 10.7 cm. were available from a Canadian manufacturer. The design of receivers for this frequency was better understood by this time. Accordingly the development of an MEW equipment was begun in the Radio Branch in July 1942. According to the "War History" it was recognized that it might have been better to adopt a somewhat longer wavelength such as 20 cm. for this apparatus, but because magnetrons, local oscillator tubes, and standard microwave "plumbing" were then available only for the 10.7 cm. wavelength, any other was out of the question.

For the first few months the Branch worked on the MEW as an internal project, but in the spring of 1943 German submarines began to operate in the Gulf of St. Lawrence, and these could not be detected at adequate ranges by the longer-wave sets such as CD and LREW. The Royal Canadian Air Force therefore placed an order for seven sets—subsequently raised to twelve—to be made in the model shop at the Radio Branch. There is surprisingly little information in the minutes of the Co-ordinating and Management Committee—indeed almost none—about the project, but we are informed by the "War History" that even this crash programme was considered too slow, so that "a hastily assembled and admittedly makeshift set was rushed to completion and, despite great difficulties, installed at Fox River, Quebec."[8] Unlike the later sets, this one had two paraboloidal antennas, one for transmitting, the other for receiving. Its usefulness was diminished by the absence of any way of identifying small friendly vessels, of which there were, of course, a great number in the Gulf, other than to have them transmit, a dangerous operation with submarines about.

One version of MEW was called MEW A/S (anti-submarine). Its antenna was designed to provide only a horizontal beam, while another MEW radar intended to detect aircraft provided coverage up to 20 degrees in elevation.

The MEW A/S sets were installed at Cape Ray, Newfoundland; Fox River, Dorval, and Bagotville, Quebec; Tofino, British Columbia; and Clinton, Ontario; the Clinton and Dorval installations were for training operators. These two, as well as the original MEW at the Radio Field Station, were used after the war in the "Stormy Weather

_____
[8] Ibid., pp. 101-102.

Project" of the Canadian Army Operational Research Group (C.A.O.R.G.) to detect falling precipitation, which gives echoes on microwave radar, greatly enlarging the range of vision of meteorological observers—a technique now widely employed. It was also expected that radar equipment of this general sort would be useful in the control of civil airports and airways, and in 1945 the R.C.A.F. assigned one set to Trans-Canada Airlines (now Air Canada). This was installed at Stevenson Field, Winnipeg, Manitoba. The new generation of airmen and even air travellers will find it difficult to imagine a time when civil aviation operated without the help of radar.

Although the MEW was distinctly a Radio Branch project, liaison was maintained with the Radiation Laboratory at M.I.T., where a similar development was under way. We are told in the "War History" that the acronym "MEW" was suggested by an engineer at the N.R.C. and adopted in both Canada and the United States. Both versions were very similar in principle and in general design, but the Canadian MEW was distinguished by the use of a slotted waveguide as a radiating element, a development due to W. H. Watson of McGill University.[9]

The external appearance of MEW A/S is shown in Figure 9.3, as it was seen in a late stage of construction. The reflector portion of the antenna was a parabolic cylinder 32 feet long with a vertical chord of 8 feet. Fed by a tapered linear array this produced a beam 1 degree wide horizontally and 4 degrees vertically. The transmitter, receiver, and monitoring apparatus were in a hut behind the reflector, both reflector and hut being rotatable at any speed up to 6 r.p.m. Locating the transmitter and receiver close to the antenna in this way not only reduced losses but obviated a good deal of complex microwave plumbing. The control circuits and the displays were in a nearby fixed building on the ground, and the operator only entered the rotatable hut for initial adjustments and for maintenance.

For the detection of high-flying aircraft the output of the transmitter could be switched to a second, smaller rotating antenna that gave a beam 1.5 degrees wide horizontally and 20 degrees vertically. This antenna was tilted upwards at an angle of 10 degrees. For this service the main beam was set at an elevation of 1 degree. For the detection of submarines the main beam was projected horizontally and the second antenna was not required.

It was found that aircraft could be detected at ranges of at least 120 miles at elevations up to 4 degrees by the main beam, while the smaller antenna covered elevations up to 20 degrees and gave a range of 60 miles.

---

[9] W. H. Watson, "Resonant Slots," *Inst. Elec. Engrs. Journal*, vol. 93. part IIIA, no. 1 (1946), p. 67.

FIGURE 9.3
MEW/AS radar

Many people were involved in this project. The report PRA-92, "MEW Antisubmarine—Book A," describing it was written by H. A. Ferris and D. W. R. McKinley.

9.6 Microwave Height-Finder (MHF)

The beam of MEW was narrow in the horizontal direction and wider in the vertical so as not to miss "seeing" any aircraft within range. As it is very important to know the height of an approaching aircraft there was a requirement for a microwave height-finder. The most obvious solution to this problem was to provide a radar equipment that would furnish a beam that is narrow in the vertical direction and could be tilted mechanically.

The origin of the MHF project in the Radio Branch was to provide a substitute for a longer-wave height-finder of British design, about 200 of which had been ordered from R.E.L. by the Royal Air Force early in 1943. Developments in the technology of microwaves led the R.A.F. to cancel this order. The R.C.A.F. had also placed an order for twenty-nine sets, and when the R.A.F. order was cancelled the National Research Council was asked to co-operate. It was clear that much of the microwave equipment designed for MEW could be used, and the R.C.A.F. at first proposed to assign a mechanical engineer to work at the Branch and do the mechanical designing, but eventually the Branch had to construct a complete prototype equipment at the field station and to send several engineers to R.E.L. to get production started there.

The operational requirements were that the equipment should find the height of an aircraft within a radius of 60 miles and at any elevation up to 30 degrees. The angular elevation of the aircraft was to be measured to plus or minus one-quarter of a degree. As the MHF was intended as an auxiliary to an early-warning radar such as MEW which provided only range and bearing, this information was sent to the operator of the MHF, who turned his antenna to the indicated azimuth and determined the angular elevation of the target aircraft.

The electronic equipment was very similar to that of MEW except for the display. The major design problem was furnished by the antenna, and the result is shown in Figure 9.4. The parabolic cylinder is 18 feet long, with a chord of 4 feet. This could be tilted up to 30 degrees about a horizontal axis, and the whole assembly, including the hut enclosing the transmitter and receiver, could be rotated in azimuth at any desired speed up to 4 r.p.m., or set in any azimuth with an accuracy of plus or minus 0.2 degrees. The parabolic cylinder could be scanned at any rate between four and twelve cycles per minute, and stopped at any desired setting.

FIGURE 9.4
MHF radar

The main difficulty encountered in the mechanical design was connected with the rotation about the vertical axis. Previous radar apparatus needing such rotation for the operation of a PPI display had merely required that it should be continuous. The MHF antenna had to be designed so that it could be stopped at any desired azimuth and remain steady even in a gusty wind.

The electrical problem concerned the translation of data on angular elevation and range into the height of the target aircraft. Besides the PPI tube there was a height-position-indicator (HPI) tube which showed a rectangular grid with range as the x co-ordinate and elevation as the y co-ordinate. After several ideas had been tried, an electromechanical computer was designed,

> consisting of a rotatable drum with height and range curves drawn on it. By pressing in, and simultaneously rotating, the drum control knob, the operator causes an electronic angle marker line to appear on the HPI, which he then varies until it bisects the echo under observation. The height of the aircraft can then be read directly in feet from the drum.[10]

This radar was apparently given a relatively low priority at R.E.L., and it was the autumn of 1944 by the time a prototype was set up at the Scarborough field station. The military situation had altered by this time to such an extent that the R.C.A.F. reduced their order from twenty-nine sets to three, apart from the original equipment built at the field station near Ottawa.

In spite of being a litle late for the war, MHF proved excellent for the "Stormy Weather" project, for it was able to measure the height of cloud layers and to detect falling precipitation, and in fact to provide vertical cross-sections of the atmosphere in any direction from a station.

J. H. Bell was in charge of the MHF project, and other names mentioned in the minutes of the C. and M. Committee are H. E. Ferris, W. J. Henderson, and D. W. R. McKinley.

9.7 Microwave Air Warning Radar (MAW)

Towards the end of the war another project was undertaken by the Radio Branch as the result of a visit to the military operations in New Guinea, Ceylon, and India by D. W. R. McKinley. The purpose of this journey, made during the first three months of 1944, was threefold:

> First, to ascertain the defects of the existing radar equipment already in the field; second, to see what could be done or was being done, ad hoc, to this equipment to make it more suitable to tropical conditions; and third, to determine what specifications were reasonable and desirable for the

---

[10] "War History," p. 105.

design of new radar sets for use in these tropical regions, taking into account both climatic and operational factors.[11]

The third objective was the main one, and McKinley discussed the requirements with the appropriate officers of the American Army in New Guinea, the Australian forces, and the R.A.F., the last two in India and Ceylon.

As a result of these discussions work began in the autumn of 1944 on a long-range early-warning radar equipment, with height-finding gear, which could be transported by air. A working model was demonstrated in January 1945, and the R.A.F. asked for five sets, hand-made by the Radio Branch, to be delivered to Burma by August of that year. The set was to go into production at R.E.L. But the war came to an end and the project was discontinued.

9.8 Very-High-Frequency Standards and Portable Precision Wavemeters

In the calibration of any set of radio apparatus it is essential to make use of a wavemeter to measure frequency. In its turn, the calibration of such a wavemeter requires an accurate instrument that will furnish a number of known frequencies. In March 1942 the R.C.A.F. asked the Radio Branch to develop apparatus of these two kinds.

The standard of frequency was to have a range of 50 MHz to 600 MHz, with calibration points every 2 MHz, specially identified at each 10 MHz and 50 MHz point. We shall not go into the technical details except to say that the frequencies were derived by using the harmonics of one temperature-controlled 10 MHz crystal, together with a multivibrator to provide the 2 MHz points.

At first the R.C.A.F. ordered three of these standards, which were delivered at intervals in 1942 and 1943. Eight more were made by August 1944 and distributed at R.C.A.F. test and development establishments. Two more were built for the National Research Council's use in its Standard Frequencies Service.

The wavemeter had to cover the range 150 MHz to 210 MHz with a reset precision of plus or minus 0.05 per cent and a long-term stability of plus or minus 0.1 per cent. It also had to be almost independent of changes in the ambient temperature. Two prototype models were built by the Branch, and after these had been approved, three wavemeters were built as a matter of urgency for the R.C.A.F., who then had fourteen more built to the N.R.C. specification by the Canadian Westinghouse Company.

"It is of interest to note," says the "War History," "that the original design [of tuned circuit] was later incorporated, without

---

[11] Ibid., p. 106.

alterations, in a wavemeter being built by R.E.L. for use in conjunction with other IFF equipment. One hundred and sixty-five of these
wavemeters were produced."[12]

9.9 The Standard-Frequency Laboratory[13]

A primary standard of frequency is one of several physical standards
which the National Research Council is required by law to maintain.
Although this requirement was set out in the National Research
Council Act, in the pre-war years the Council had never actually
owned any apparatus for the maintenance of a standard of radio
frequency. What apparatus there was at the Sussex Street building
was the property of the Canadian Radio Broadcasting Commission
(later the C.B.C.) until 1938, when an Order-in-Council caused it to be
transferred to the Department of Transport to be used for monitoring
the frequencies of radio transmissions.

It was rightly considered intolerable that Canada should be without a satisfactory national stanard that could provide services similar
to those rendered by the National Physical Laboratory in England and
the Bureau of Standards in the United States. In 1939 a 50 KHz
temperature-controlled crystal was built into apparatus that furnished certain standard frequencies by telephone line to various
laboratories that had need of it. It was also used to calibrate apparatus
for the Armed Services and other laboratories. However, there was no
time during the war to establish a real primary standard of frequency
that would be independent of any other country, although towards
the end of the war more and better apparatus was obtained, and the
frequencies were related to the astronomical observations made at the
Dominion Observatory.

For use at the radio field station the 50 KHz signal from the crystal
was multiplied to 2 MHz and broadcast from the Sussex Street laboratory. This signal had "a precision of one part in ten million and [was]
modulated by a 1-kilocycle tone of the same accuracy."[14] Anyone in
the Ottawa area could pick up the signal if he needed an accurate
2 MHz radio frequency or a 1000 Hz tone.

---

[12] Ibid., p. 113.

[13] This project is dealt with in the "War History" under "Projects for the Air
Force." There seems to be no particular reason for this, but I have followed the
precedent.

[14] N.R.C. Review (1948), p. 137.

Chapter 10

RADIO FOR CIVIL AVIATION

In 1944 and 1945 three conferences on radio for civil aviation were held at the instance of the British government. At the first of these, held in London in February 1944,[1] it was recommended "that the Canadian Government, with such technical assistance as can be made available from the United Kingdom and elsewhere, should be invited to give the best demonstration of a full-scale working model of the radar range and associated radar systems that their other commitments permit." The Canadian delegation to this conference consisted of four people: Dr. A. G. Shenstone (at that time the N.R.C. Liaison Officer in London); Group Captain H. B. Godwin, Director of Signals, R.C.A.F.; W. A. Rush, Controller of Radio in the Department of Transport; and S. S. Stevens, Communications Superintendent of Trans-Canada Air Lines.[2] It is worth noting that Stevens was the only representative of an airline at the conference, which reviewed the entire field of radio as applied to civil aviation. His presence in London was undoubtedly due to the fact that in the latter part of 1943 Trans-Canada Air Lines had been begging for information about radar, for they suspected that their American counterparts were being given information of this sort unofficially.[3] Stevens' attendance at the conference must have given him the information that he wanted.

When the delegates, except for Shenstone, came back to Canada they got to work at once. J. E. Clegg of the Telecommunications Research Establishment also came to Canada, and on March 15, 1944, he was at a meeting in Mackenzie's office with Rush and C. J. Acton of the Department of Transport, Group Captain Godwin, and Stevens. It

---

[1] It was known as the Commonwealth and Empire Conference on Radio for Civil Aviation (C.E.R.C.A.).

[2] Now Air Canada.

[3] Mackenzie, "War Diary," November 26, 1943.

was pointed out that the Radio Branch was very short of engineers because of the pressure of war work, and could not spare more than two men. Later it was decided that Professor Arnold Pitt of the Physics Department of the University of Toronto would be asked to supervise the design of the radar range, assisted by engineers from the N.R.C., the Department of Transport, and the Bell Telephone Company. The R.C.A.F. supplied an aircraft and a pilot.

The purposes of the radar range system are three: (1) to make the distance of the aircraft from a given station available to the pilot at every moment; (2) to enable the pilot to determine his position relative to the track he is supposed to be flying; and (3) to give him a means of homing on to the station. The equipment consisted of two parts: a radar set fitted into the aircraft, and a radar range on the ground.

Short pulses emitted from the airborne radar set at a frequency of 248 MHz were picked up by the radar range receiver. Each pulse so received caused two pulses to be sent out from the ground station at 294 MHz. One pulse was radiated to the left of the beacon, the other, delayed by 10 microseconds, to the right. If the two pulses were of equal amplitude on the indicator in the aircraft, it was on the beam. At intervals the station identified itself in Morse code, which appeared visually on the indicating tube.

The first radar range beacon was installed near Malton airport, west of Toronto, in August 1944. After some improvements had been made, the equipment was demonstrated to the Second C.E.R.C.A. Conference, held at Ottawa in November 1944. For this purpose a second station was established at Uplands Airport, Ottawa.

It was agreed that while this system worked, it was unsatisfactory for two reasons: first, because there was too much information displayed on the cathode-ray tube indicator, and secondly (and more important), because the relative amplitude of the two pulses could be affected by reflections from objects on the ground. To get rid of the second of these objections another system called the Modified "H" system was proposed. This employs two beacon stations a few miles apart. The principle is that if the two pulses reach the aircraft at the same time, it must be on the right bisector of the line joining the two beacons. This system was demonstrated at Uplands and was fairly successful. Nevertheless, the Branch did not pursue this further, but concentrated on an airborne distance indicator "to provide the pilot of an aircraft his distance from one or more ground points, continuously on a meter with an accuracy of ± 1 mile, up to a distance of 100 miles."[4] The presentation of the distance on a meter was an important advantage. This apparatus consisted of a ground station which

---

[4] N.R.C. Radio Branch, *Airborne Distance Indicator* (Ottawa), July 15, 1945.

emitted 5-microsecond pulses at a frequency of 202 MHz when inter-rogated by 222 MHz pulses from a transmitter in the aircraft, and a receiver in the aircraft which translated the delay in the received pulses into meter readings. H. Ferris was responsible for this ingeni-ous circuit.

In May 1945 a complete system was installed at the experimental station of the U.S. Civil Aeronautics Authority near Indianapolis, Indiana. After being demonstrated to numerous interested parties it was left there, although without proper maintenance, until De-cember. Another was flown to England in July so that it could be demonstrated at the Third C.E.R.C.A. Conference. One of the recom-mendations of this conference reads as follows: "Distance Indication—Pending the deliberations of a special post-conference Panel the system proposed by Canada for distance indication in the aircraft was recommended for international standardization."[5]

In January 1945 two of these systems were installed near Washington, D.C. for direct comparison with American systems operating on a frequency of 1000 MHz, installed at the same sites. These comparisons seemed to show a clear superiority for the N.R.C. distance indicator.[6] Nevertheless, the United States unilaterally standardized on a frequency of 1000 MHz for this purpose.

Improvements of various kinds were made to the radar distance indicator during 1946, and a number of experimental units were built so that the Department of Transport and Trans-Canada Air Lines could use them under service conditions.

The section of the Branch that dealt with this subject was also charged with the task of adapting radar to survey operations, a matter of great importance to Canada with vast territories to be mapped. Two problems were being investigated: the problem of measuring long baselines with good accuracy, and the development of a recording radar altimeter with a narrow beam, so that the altitude of a photo-graphic survey aircraft above a particular point on the ground could be known exactly. In subsequent years these researches yielded enormous dividends in the mapping of Canada and of other countries as well.

---

[5] *Third Commonwealth and Empire Conference on Radio for Civil Aviation: Record of the Conference and Demonstrations, July-August, 1945* (London: H.M.S.O., 1945), p. 44.

[6] N.R.C., Electrical Engineering and Radio Branch, Report PRA-133, "Tests of NRC 200 Mc Distance Indicator at Washington, D.C., January 1946," March 1946.

Chapter 11

TRANSITION TO PEACE

Those who passed the years 1939 to 1945 in Canada will remember how the pessimism of 1940 and 1941 gradually changed to hope and then confidence with victories over the Axis powers in various theatres of war in 1943. The rate at which this process took place depended on individual temperament. C. J. Mackenzie was certainly among the most optimistic, and appears to have communicated his outlook to the Honorary Advisory Council, for as early as December 17, 1941, the Council asked him to appoint a special committee to consider the probable trend of post-war research in the Laboratories.[1] This committee was to report from time to time.

During the next three years Mackenzie, as opportunity offered, gradually gathered ideas from both inside and outside Canada on the nature and scope of the research that should be carried on in the Laboratories after the war was won.

Late in 1944 General A. G. L. McNaughton decided that he did not want to return to the presidency, and on October 13 Prime Minister W. L. M. King appointed Mackenzie President. At the same time he announced the appointment of C. D. Howe as Minister of Reconstruction and Chairman of the Privy Council Committee on Scientific and Industrial Research—the cabinet committee to which the Honorary Advisory Council reports. Howe and Mackenzie were great friends, were both engineers, and felt the same way about what should happen to the Laboratories. In Mackenzie's next budgetary submission to the Treasury Board, he proposed to maintain or increase the scale of expenditures for the National Research Council, to set up a military research organization (this was to lead to the Defence Research Board in 1947),[2] to expand the nuclear research programmes

---

[1] Proceedings, 137th meeting, minute 16.

[2] See Wilfrid Eggleston, *National Research in Canada: The NRC 1916-1966* (Toronto: Clarke, Irwin & Co., 1978), pp. 271-75.

for the peaceful uses of atomic energy, to expand medical research, and to establish several new divisions, among them a Division of Radio and Electrical Engineering.[3] Though the government accepted all these policies in principle, it was to be the end of 1946 before the Radio Branch became part of a new Division of Radio and Electrical Engineering although, as we saw in Chapter 6, the Electrical Engineering Section was joined to the Radio Branch on January 1, 1946, to become the Electrical Engineering and Radio Branch of the Division of Physics.

While some of the professional staff left for other employment, there was no marked depopulation after the war. For the year 1946 the N.R.C. Review lists 57 of these, 100 technical personnel, 20 administrative staff, and 53 "employees on prevailing rates."[4] The Branch was organized in the following sections at that period: Radio Aids to Air Navigation, headed by D. W. R. McKinley; Radio Aids to Marine Navigation, H. R. Smyth; Microwave, G. A. Miller; General Electronic Application, J. E. Breeze; Electrical Engineering, T. W. Mouat; Engineering Design, H. E. Parsons; Instruments, C. F. Pattenson; Model Shops, I. L. Newton; Reports and Publications, J. W. F. Chisholm.

After the capitulation of the Germans and the Japanese it became possible to satisfy some of the considerable public curiosity about secret wartime developments. In the case of radar it became advisable to do so, in view of extravagant publicity about its possible peacetime applications to civil aviation.

An exhibition of radar apparatus was therefore assembled by the Radio Branch in co-operation with the Armed Services, Research Enterprises, Ltd., and the National Film Board. It was shown during the week of November 19, 1945, in the electrical engineering laboratory in the Sussex Street building, with a film in the auditorium. Unfortunately the limited size of the laboratory (the largest one available) made it impossible to open the exhibition to the general public, and invitations were therefore issued to appropriate groups, including the Chiefs of Staff of the three Services, the members of the House of Commons, the Press Gallery, members of professional and scientific societies, the universities, industry, N.R.C. staff, and others. Twenty-six showings were given, each preceded by a film that gave a general introduction to radar. Complete radar sets, demonstration models, and samples of equipment were arranged in the electrical engineering laboratory, with demonstrators to answer questions. Some of the larger radars such as GL Mark IIIC and MZPI were drawn up outside the building and placed in operating condition. More than

---

[3] See also my Physics at the National Research Council of Canada, 1929-1952, pp. 125-26.

[4] Compare Figure 6.1, p. 29.

2,000 people saw the exhibition, which was considered a great success.

During this period the Branch was the scene of part of a spy story that in its entirety involved a number of scientists in the public service. This conspiracy, which attracted intense public attention when it became known in 1946, might never have come to light except for a crisis of conscience on the part of a clerk in the Soviet embassy called Igor Gouzenko, who revealed the details of the plot to the Canadian Government and asked for political asylum.

Only two of the staff of the Radio Branch were involved: Edward W. Mazerall and Durnford Smith, both born in Canada of Canadian parents. Our information about their activities is derived from the report of the Royal Commission that was appointed to investigate the matter, a document of more than 700 pages.[5] The commissioners were the Honourable Justice Robert Taschereau and the Honourable Justice R. L. Kellock. The seriousness with which the matter was treated is evident from the comparatively short time that elapsed between the Order-in-Council and the completion of the report.

Neither Mazerall nor Smith were approached directly by anyone from the Soviet embassy, although it is clear that their political leanings were known and that the embassy also knew where they worked. The contacts were made through one D. G. Lunan, who at the time was editor of a monthly pamphlet for soldiers, *Canadian Affairs*. Lunan, who was a member of one of the quasi-intellectual groups that proliferated at that time as communist front organizations, was persuaded to try to subvert a group consisting of Smith, Mazerall, and another man who was with the Department of National Defence at Valcartier, Quebec.

Lunan succeeded in getting these men to obtain secret information and pass it to the embassy. Smith was given the code name *Badeau*, Mazerall was *Bagley*. To make a long story short, Durnford Smith gave a good deal of secret information, in particular about the CB radar.[6] There is evidence that he turned over more than 700 pages of documents in all. Mazerall provided only some information regarding the development of radio aids to air navigation.

The Soviet embassy was prepared to pay small amounts of money for this information—the derisory sum of $30 to each man is mentioned—but there is no firm evidence that either Smith or Mazerall took it, and they flatly denied doing so. In the words of the Royal

[5] *The Report of the Royal Commission appointed under Order in Council P.C.411 of February 5, 1946 to investigate the facts relating to and the circumstances surrounding the communication, by public officials and other persons in positions of trust, of secret and confidential information to agents of a foreign power* (Ottawa: King's Printer, June 27, 1946).

[6] See p. 93 above.

Commission, "Whatever may be the truth, it seems sure that even if money were given, it was not this motive that prompted Lunan and his group to act as they did. The motive of working for the Soviet regime and the Communist cause was undoubtedly the primary factor."[7] An explanation of the actions of these men is a problem for the psychologists. I shall leave it at that. It is, however, a curious fact, as D. J. C. Phillipson has pointed out,[8] that of a dozen members of the professional staff who were asked about this "Gouzenko affair" thirty years later, few could remember being either surprised or much interested in it in 1946—at that time few scientific workers were giving much thought to politics; they were too absorbed in their work. I began to work in the Sussex Street Laboratories in June of that year and I cannot recall any distinct impression or disturbance resulting from these disclosures.

In the trials that followed the enquiry, Mazerall was sentenced to four years in prison, Smith to five years.

It remains to summarize in the final chapter the work of the Electrical Engineering and Radio Branch during the year 1946, that is to say, until it became an autonomous division of the National Research Council.

---

[7] *Report of the Royal Commission*, p. 160.
[8] Communication to Dr. A. W. Tickner, July 11, 1979.

Chapter 12

THE BRANCH IN 1946

12.1 Introduction

On December 15, 1945, B. G. Ballard found himself the Officer-in-Charge of an entity called the Electrical Engineering and Radio Branch of the Division of Physics. That was its official title, but in reality it would not have made any difference to its programme if the Division of Physics had not existed. The Branch lasted just over a year, and then became the Division of Radio and Electrical Engineering, with Ballard as Director. Ballard was not a radio man but an electrical engineer of an older school, a man of considerable competence in various fields who had come to the National Research Council in 1930 from the Westinghouse Company,[1] and therefore had been considerably longer in the service of the Council than any of his staff.

Ballard's first preoccupation was with the question of laboratory space for his Branch. This was not only inadequate but grotesquely inconvenient, for the work in radio had to be divided between cramped space on the third floor of the Sussex Street building and the temporary buildings at the field station. These conditions, Ballard wrote, were

> not only unsatisfactory but in many cases even hazardous. In particular, a substantial portion of the laboratory is located in temporary frame structures and the fire risk to life and to valuable equipment is a cause of concern despite the many precautions which have been taken. Furthermore, the separation of the Branch into two isolated groups more than nine miles apart introduces administrative and operational difficulties which the proposed quarters will rectify. The excellence of the work

---

[1] For his pre-war activities see Middleton, *Physics at the National Research Council of Canada, 1929-1952*, pp. 29-33.

which has been completed under these conditions is a tribute to the staff.[2]

This is the first public reference to the "proposed quarters," which eventually became the substantial building known as M-50 on the Montreal Road site, occupied in October 1953. C. J. Mackenzie had no doubt about the necessity for such a building, the more so because the overcrowding in the Sussex Street building was hampering the work of the rest of the Division of Physics.

As previously noted, Research Enterprises, Ltd. ceased to exist in 1946. The company had a field station on the cliffs at Scarborough, Ontario, overlooking the lake, and the National Research Council acquired the site for the Branch to use in the study of radar problems over water.

There was a certain amount of dislocation caused by resignations after the war, but less than might have been expected. However, "requests to undertake new projects [are] well in excess of the capacity of the branch to deal with them and many interesting problems are rejected regretfully."[3] In the list of staff for 1946 we find fifty-seven professionals. It is noteworthy that all but five of these were graduates of Canadian universities.

In the remainder of this chapter the chief researches of the Branch will be ascribed to the section that co-ordinated the work, following the arrangement in the *N.R.C. Review for 1946*, where it is noted that many of these projects represent the co-operative effort of more than one section. The work of the Radio Aids to Air Navigation Section during the year has already been dealt with in Chapter 10.

12.2 Radio Aids to Marine Navigation

The Marine Section under H. Ross Smyth was left with a ready-made chore at the war's end because of the existence of a surplus of the very successful type 268 marine radar sets, originally designed for motor torpedo boats.[4] These had proved to be suitable for installation in merchant ships. Accordingly much of the effort of the Section in 1946 was in the installation of these surplus radar sets on Canadian ships. (It is salutary to remember that Canada once had a merchant fleet.) "Many ship's captains," says the *Review*, "have reported in glowing terms of radar performance."[5] Nevertheless, it was clear that an equipment designed for small naval vessels, however successful, should be modified for civil use, especially in the direction of making

---

[2] *N.R.C. Review for 1946*, p. 77.

[3] Ibid.

[4] See p. 53 above.

[5] *N.R.C. Review for 1946*, p. 79.

it less complex. The Section began to develop a set for merchant-marine use incorporating the latest techniques.

It was also evident that an adequate radar set mounted on land at a suitable site near a harbour would be of great use in controlling the traffic during fog, rain, or snow, and therefore in preventing collisions in bad weather. During 1946 the Section developed a harbour-control radar and at the end of the year this was being installed at Camperdown, Nova Scotia. For the same station a high-frequency CRDF equipment, made commercially to National Research Council specifications, was modified in the Section so that it would operate on the much lower frequencies used in marine communications.

In the autumn of 1945 a sixty-five-foot vessel with a 100-horsepower diesel engine was purchased for the Section and christened *Radel*. During the following winter and spring it was modified to suit the requirements of the laboratory.

The other main sort of activity in 1946 was the analysis of data obtained previously at the request of the Canadian Radio-Wave Propagation Committee, concerning the propagation of K-band and S-band radiation over land.[6] This analysis provided some very interesting results that were later useful in the theory of radio-wave propagation in the lower atmosphere. Ionosphere studies also received attention; the Section began the development of a completely automatic recorder that was to sweep over the radio spectrum from 2 MHz to 30 MHz each thirty seconds.

12.3 Microwave Section

During the war the physicists and engineers of the Microwave Section, headed by F. H. Sanders and later W. J. Henderson, had developed or learned a great deal of technique and constructed much equipment in this field. The cessation of hostilities made it possible to begin to apply this new knowledge to more fundamental researches. During the year 1946 the professional staff of the Section consisted of three physicists and eight engineers, with G. A. Miller as Section head.

The year 1946 saw the beginning of a long series of researches in radio astronomy, which was the particular interest of A. E. Covington. In working with 10-cm. waves at the Bell Telephone Laboratories during the war it had been found that the sun was a strong emitter in this region of the electromagnetic spectrum, and in 1946 a great deal of effort was put into the design and construction of instruments for measuring this "solar noise." The first four-foot radio telescope was assembled from various components used in various radar sets such

[6] See p. 62 above.

as GL IIIC, 268, and MEW. During the last eight months of the year a recording radiometer was kept in service nearly all the time. On November 23 there was a partial eclipse of the sun. From the records obtained during this eclipse it was found that while 10-cm.-band noise came from all parts of the solar disc, it was particularly strong from a notable group of sunspots.[7]

Work with very short waves requires special vacuum tubes. During the last months of the war a vacuum-tube laboratory was organized in the Branch, and in 1946 this laboratory devoted much of its time to the development of a klystron for the generation of 8-mm. waves.

The Section did a great deal of work on the design of antennas to be installed in high-speed aircraft, necessarily within the aircraft in order to minimize air resistance. It also provided expert assistance in making a number of such installations.

In a country like Canada radio communication is of very great importance, especially in the far north. It was suggested to the Branch by a Montreal firm that "very high frequency" radio—say in the region of 75 MHz—might be useful in the north; so the Microwave Section set up a radio link of this sort between Ottawa and Montreal in order to investigate the possibility of using this region of the spectrum for such purposes. The reader may note that this frequency is just below the band now used for frequency-modulation broadcasting.

In a related development, but at the instance of the railways, the laboratory helped to set up a chain of 10-cm. radio stations between Windsor, Ontario, and Montreal. This was the precursor of the many microwave circuits that today carry every sort of communications from end to end of the country and indeed over most of the land areas in the world.

12.4 Electronics for War and Peace

The Army Section, headed by W. Happe, Jr. during the war, changed both its name and its leadership in 1946. J. E. Breeze became Section head, and the laboratory was named the General Electronic Application Section. Most of the army engineers who had been attached to the Section during the war had been given their discharge or had been posted elsewhere. Only six professionals, all engineers, remained in 1946.

Nevertheless, the Section continued to co-operate with the Canadian Army. The main projects involved, namely, the MZPI and the CB, have already been dealt with in Chapter 8. These indeed took up most of the time of the reduced staff, but there were some developments of general interest.

---

[7] A. E. Covington, *Nature* 159 (1947), 405-406; also N.R.C. Report ERA-216, 1952.

One of these, which arose from the military work, was a study of regulated direct-current power supplies, valuable in a wide variety of electrical problems. These had been designed *ad hoc* as the need arose. Three papers on various aspects of this technique were prepared during the year. Another related study concerned the design of high-frequency, high-voltage power supplies.

At the request of the Division of Applied Biology the laboratory developed an instrument for the automatic control of pH (hydrogen-ion concentration) in fermenting solutions. This involved the measurement and control of the very small voltages generated at the glass electrodes used in such measurements, and provided an interesting problem in circuit design. The same project led to the development of a metering pump, controlled by the instrument that measured the pH, which added an alkaline liquid to the fermenting solution a drop at a time.

In the tradition of co-operation with standardizing bodies, always a feature of the National Research Council, the Section correlated electronic standards for the relevant Inter-Service Committee.

12.5 Electrical Engineering

The Electrical Engineering Section was installed, as it had been since 1932, in a large two-storey room at the back of the building in Sussex Drive. In 1946 it was the largest section of the Branch, employing fourteen engineers and headed by T. W. Mouat.

Just before the war the most visible activity of this laboratory had been in the field of high-voltage power engineering, but this work had been entirely abandoned during the war in favour of more urgent demands. In 1946 the 1,000,000-volt impulse generator, built in 1938, was restored to operation and its associated oscillograph much improved. The laboratory was able to announce that the capacity of the generator could be extended to 2,000,000 volts with available components, if this should be required.

The ability of this Section to deal with high voltages was put to use on behalf of the Atomic Energy Division. In 1946 a 5,000,000-volt direct-current generator was designed and under construction; this had to be sealed within a steel vessel filled with air at high pressure in order to prevent flashovers by providing a high dielectric strength.

Although the Electrical Engineering Section was not concerned with radar, some radar techniques spilled over into other fields. An example of this is an instrument for locating faults in long transmission lines used for the transmission of power. If a short pulse is imposed on the line it will be reflected from any discontinuity such as an open circuit, a short circuit, or even a partial fault, and a signal will be returned on the line. The time required for the signal to return indicates the location of the fault, and the nature of the trouble is indicated by the wave-form of the reflected pulse. Such an instrument was constructed and applied in 1946.

The end of the war saw the start of a good deal of industrial activity in the form of redesign of products that had fallen behind the state of technique. For reasons of safety or other reasons, some of these products were subject to government regulation. An example was the oil burner for domestic heating. The Canadian Standards Association had an agreement with the National Research Council for the type-testing of such burners, and this field expanded so rapidly that in 1946 the Electrical Engineering Section had to examine and test more than twenty-five types of oil-burning equipment.

12.6 Various Services

Because of the way in which this book is organized, not enough attention has been paid to three sections of the Branch without which it could not have operated at all. These are the Engineering Design Section, the Instrument Section, and the Model Shops Section.

Throughout the war and afterwards the Engineering Design Section under H. E. Parsons was an essential link in a chain that began with the ideas of the other sections and ended in the production of useful equipment. In this context "engineering" must be interpreted as mechanical engineering. Physicists and even electrical engineers are seldom competent in mechanical design, and the availability of a good design team was essential at all times. The Section, in fact, "contributed to practically every project undertaken by the Branch,"[8] as Parsons was proud to state. Two of the most critical fields of its activity were antenna structures and the mechanisms by which antennas were rotated. The Section was obliged "to translate a rough model or a computation into a suitable mechanical design embracing a range of sizes from a 3/4 by 12 in. polystyrene tube for an airborne antenna to a 30 ft. parabolic reflector."[9] The rotators came in a similar range of sizes, and always had to be designed to resist all kinds of weather and extremes of ambient temperature.

In 1946 the thirty-foot paraboloid, intended for radio astronomy, was being designed, as was the mechanism to rotate it in altitude and azimuth. Much time was devoted to an "axis converter" to control the variable-speed motors of this mechanism so that the big antenna would follow any desired star across the sky.

Another related instrument that was almost completed during the year was a mechanical device that was fed range and azimuth from

---

[8] N.R.C. *Review for 1946*, p. 83.
[9] Ibid.

the CB radar and delivered the coresponding map co-ordinates. The Engineering Design Section also had charge of the drafting room.

The Instrument Section was concerned with the maintenance and calibration of all the many kinds of electrical measuring instruments in use in the Branch. During the war it had been impossible to do this systematically, so that in 1946 the Section was very busy getting things in order. Apart from this purely service function it engaged in the development of apparatus for the measurement of electrical "noise," at the request of the Canadian Standards Association.

We have noted several times that in the course of the war the so-called "model shops" of the Branch were obliged to do a great deal of manufacturing in order to satisfy urgent requirements of the Armed Services. As a result they were operated beyond their normal capacity, so that it was impossible to maintain the machine tools properly or to replace those that wore out. In 1946, although the shops were still very busy, it was possible to begin a programme of maintenance, repair, and replacement, and to renovate electric wiring in the interests of safety. This was carried on in a somewhat tentative manner in view of the fact that a new building for the Branch was being designed.

One advance made in 1946 was to set up a definite establishment for the shops, a move which led to an increase in morale and satisfaction among the skilled workmen upon whom, after all, the success of the Branch had to depend.

Chapter 13

EPILOGUE

The developments described in this book, and especially the microwave radar sets, must be thought of as the very beginnings of a new technique, comparable with the locomotive of Robert Stevenson or the flying machine of the brothers Wright. It is no wonder that the results sometimes look primitive, in fact, unsatisfactory to an aesthetically-minded engineer. The one overriding requirement was to get something that would work, and get it as quickly as possible. As the war progressed the designs became more refined, and in one instance, the antenna of the type 268 radar (Figure 7.4), elegant. Some equipments completed soon after the period covered by this book were, quite naturally, more sophisticated.

The electrical circuits of most of these radar sets had to deal with fairly large amounts of power. They had therefore to be somewhat bulky. The standard module in the industry was the so-called "rack panel," nineteen inches wide and a multiple of an inch and three-quarters deep. Techniques were widely available for assembling standard components on sheet-metal "chassis" fastened behind such panels, and this is why the external appearance of such apparatus was almost always much like the examples shown in Figures 7.2 and 7.3. Sometimes limitations of space and weight posed a serious problem in these early years. The picture changed completely after the invention of the transistor and the tremendous development of solid-state devices in the 1950s and 1960s.

My reason for mentioning these matters is simply to underline the remarkable feat performed by the physicists and engineers of the Radio Branch in taking what was available and making usable pieces of equipment so quickly. We have seen that their competence and speed sometimes surprised the Americans and astonished the British. They themselves had no time to be surprised while this was going on, but the recollections of the survivors contain a measure of quiet satisfaction. Nearly all of them were graduates of Canadian universities, a fact that ought to do a little to counter our endemic self-deprecation.

The radar development was only a part of the war effort of the National Research Council, but a major and distinctive one. The contributions of other Divisions are being recounted in the other volumes of this series. The technical momentum engendered by it was maintained after the war in the Division of Radio and Electrical Engineering, and in the present Division of Electrical Engineering progress continues to this day.

Appendix A

ABBREVIATIONS USED IN THIS BOOK

Note that abbreviations for institutions are punctuated (e.g., N.D.R.C.), abbreviations for types of equipment (e.g., MAW) are not.

AI	[Airborne radar for] aircraft interception.
AIR	Automatic ionosphere recorder.
APF	Accurate position finder [radar]. A part of GL IIIC.
A.S.E.	Admiralty Signals Establishment (British).
ASV	Air-to-surface vessel [radar].
C. & M.	Co-ordinating and Management [Committee].
CB	Counter-bombardment [radar].
C.B.C.	Canadian Broadcasting Corporation
CD	Coast-defence [radar].
CDX	Coast-defence gun-laying radar.
C.E.R.C.A.	Commonwealth and Empire Conference on Radio for Civil Aviation
CH	Chain home, a British early-warning radar.
CNJ	Canadian naval jammer.
CRDF	Cathode-ray Direction Finder.
CSC	Canadian submarine-control [radar].
C.W.A.C.	Canadian Women's Army Corps.
FAW	Forward-area warning [radar].
GCI	[Radar for] Ground control of interception.
GL	Gun-laying [radar].
GL IIIC	Canadian microwave gun-laying radar.
HPI	Height-position indicator [display].
Hz	Hertz (= 1 cycle per second).
IFF	Identification, friend or foe.
K band	Wavelength of about 1.25 cm.

LREW	Long-range early-warning [radar].
MAW	Microwave air warning [radar].
MB	A British radar.
MEW	Microwave early-warning [radar].
MEW A/S	Microwave anti-submarine [radar].
MHF	Microwave height-finder [radar].
M.I.T.	Massachusetts Institute of Technology.
MZPI	Microwave zone position indicator [radar].
N.D.R.C.	National Defence Research Council (U.S.).
N.R.C.	National Research Council of Canada.
NW	"Night watchman" (a harbour-defence radar).
PPI	Plan position indicator [display].
R.A.F.	Royal Air Force.
R.C.A.F.	Royal Canadian Air Force.
R.C.N.	Royal Canadian Navy.
R.C.N.V.R.	Royal Canadian Naval Volunteer Reserve.
RDF	Range and direction finding (later called radar).
R.E.L.	Research Enterprises, Ltd.
RF	Radio frequency.
RX/C	Microwave ship-borne radar.
RX/F	A ship-borne radar (= type 268).
SA	A British naval radar.
S band	Wavelengths near 10 cm.
SLC	Searchlight-control [radar].
SS	A British Naval radar.
SS-2C	Microwave ship-borne radar (= RX/C).
SW-1C SW-2C	Ship-borne radars (See CSC).
T.C.A.	Trans-Canada Air Lines (now Air Canada).
VEB	Variable-elevation beam [radar].
X band	Wavelengths near 3 cm.
ZPI	Zone position indicator [radar].

Appendix B

SUGGESTED SPECIFICATION FOR A GL 3 SET

Record of a discussion at the National Research Council, on October 28, 1940

1. In view of the cable from England stating that the range of 14,000 yards had been obtained on 10 cm. equipment, it seemed desirable to construct a proto-type 10 cm. GL set at once but to continue work on a micropup transmitter and on a receiver for 50 cm., since this was required for CD.

2. *Performance Specification*
   The following was proposed, based on the general staff GL 2 specification:

   Range   ±50 yds. acceptable; ±25 yds. desirable; from 2,000 to 14,000 yds.
   ±250 yds. to maximum possible range for search purposes.
   Bearing   ±1/4° acceptable; ±1/6° desirable.
   Elevation   ±1/4° acceptable; ±1/6° desirable. From 8′ to 90°.

   Continuous range, bearing and elevation to be fed out on hs and ls magslip transmitters; ratio 36:1. Maximum rate of change of bearing, 10° per second.
   Provision to be made for a displacement correction to range and to set in a permanent range correction.
   Display to central zero instrument for bearing and elevation and centralized echo cathode ray tube for range. Provision to be made for target selection by stroboscope.

3. *Cabin Specification*
   The cabin and general structure should be based on the GL 2 design and should form a sufficiently rigid structure when jacked up to give the necessary stable platform for the aerial system to maintain alignment to better than 1/6′ under all wind conditions.
   Transmitter and receiver would probably occupy 3 standard racks. Provision should be made for seating 3 operators and 1 transmitter attendant. This would probably require a cabin about 9 ft. × 7 ft. × 8 ft. (high).

Heat insulation should provide for Canadian conditions—heat and ventilation should be adequate.

Provision should be made as in GL 1 for entry of electrical supplies and for connection of telephone and magslip cables.

Aerials to be demountable for transport, the design to be such that the equipment can go into operation very quickly.

4. *Aerials*

Two alternatives should be considered:

(1) One transmitter paraboloid—4 receiver paraboloids.

(2) One transmitter paraboloid and one receiver paraboloid with 4 dipoles.

The aperture of the elevation paraboloids should provisionally be 6 or 8 wave lengths to enable a minimum angle of 8' to be attained. The aperture of the horizontal paraboloids must be a compromise between giving sufficient breadth for search and obtaining the necessary angular definition—probably a 4 aperture would be satisfactory.

Horns should be mounted in a bearing similar to the Bell Laboratories Naval set design. If this is in production it might be adopted.

Drive for bearing and elevation might be by amplidyne type motors controlled by the output signal of bearing and elevation receivers. It should be possible to control these motors by hand or throw of a switch.

Bearing should be transmitted by a shaft from the turntable to the bearing scale and magslips mounted in the cabin with provision for a displacement correction to range exactly as in GL 2.

Elevation should be transmitted by hs and ls magslips to the predictor and provision should be made for setting in a correction to elevation by rotation of the magslip housing. Transmission of elevation to the elevation dial in the cabin may be by magslip.

Handwheels for range setting should be of the two-speed type as in GL 1 and GL 2, in which one revolution of the handwheel corresponds to 100 yards. Transmission to be by hs and ls magslips with provision for setting in displacement correction, exactly as in GL 2.

As a refinement, provision might be made for continuous rotation of the aerials and for provision of a radial time base for early warning if the maximum range warrants this.

5. *Transmitter*

The transmitter should use the E1189 magnetron and should aim at 20 k.w. output. To provide for future improvement, the power pack should provide for 50 k.w. The pulse width should be adjustable in 3 steps from .5 to 5 microseconds as in the Bell Lab. design. The modulator should make use of the Wembley heavy-current tube developed for this work.

The recurrence frequency should be adjustable from 500 to 1000 cycles.

6. *Receiver*

The receiver should in the first instance use the following components:

(1) Bell Lab. crystal mixer and local oscillator.

(2) Standard IF amplifier to be developed for ASV. The four receiver horns should each have a separate receiver. This will facilitate provision of the central zero instrument reading for bearing and elevation. It will also provide for control of the amplidyne circuits.

Range should be determined by centralizing the CR echo. Experiments should be carried out to determine whether the same order of accuracy can be obtained by a phasing circuit as with the potentiometer of GL type. If so, the first may be adopted.

A stroboscopic signal selector circuit with automatic gain control should be provided as in GL2. The circuit diagram is available.

The over-all sensitivity of the receiver should be as high as possible. 10^{-12} watts is desirable and 10^{-11} watts is acceptable as the first shot.

An anti CW filter circuit should be provided to counteract jamming and a long delay. CRO should be incorporated as soon as available.

7. *Summary of action required by NRC*

(1) Valves* —   Development contract to be placed by Research Enterprises with Northern Electric for E1189 and for modulator valve.

(2) Cabins —   Consult with Colonel Taber about supply and about dimensions.

(3) Aerials —   Paraboloids might be ordered from the firm supplying Loomis. Enquiry to be made from Bell Lab. as to aerial turntable. Enquiry for standard searchlight amplidynes to be made from G.E.C., Schenectady.

(4) Receiver —   All the following components from Bell Lab. —
Crystals, 1020Y valve
Bell Lab. local oscillator and crystal mixer design to be made.

(5) Transmitter —Order magnets from G.E.C., Schenectady.

(6) Generator —   Consult Colonel Taber about suitable stabilized AC generator.

(7) Enquire about production of magslips in Canada or alternatively enquire about the provision of the necessary magslips totalling 7 per set.

(8) Initiate experiments on the accuracy of the phasing method for determination of range.

(9) Preliminary experiments using rough assembly to determine maximum ranges. Desirable to have this information within one month.

(10) Draftsmen to be at once employed on the general layout of components and in particular on the mechanical arrangements.

(11) Enquiry to be made as to production of necessary gears. Gears will be required for the following functions:
30:1 and 60:1 reduction in aerial drive for bearing and elevation;
Transmission of bearing to bearing scale and to bearing magslips;
Transmission of elevation to magslips;
Drive free range handwheel to phasing circuit and to range magslips.

---

* In North American terms, vacuum tubes.

Appendix C

LIST OF REPORTS ISSUED BY THE RADIO BRANCH

Report No.	Title	Author
PRA-1	Test of Marconi Ground Radio Receiver	J. T. Henderson
PRA-2	Test of Three Materials for Coating Radio Coils	J. T. Henderson
PRA-3	International Business Machines Corporation Letter	J. T. Henderson
PRA-4	Report of Trip to Peck Television Corporation, RCA and Western Electric Company	J. T. Henderson
PRA-5	Report on Trip to England	R. W. Boyle
PRA-6	Electrical Methods of Fire Control in Great Britain	F. V. Heakes and J. T. Henderson
PRA-7	Electrical Methods of Fire Control in Great Britain (Report No. 2)	J. T. Henderson
PRA-8	Flight Tests of Cathode Ray Direction Finder for Use in Aircraft	D. W. R. McKinley
PRA-9	Measurement of Power at Ultra High Frequencies (up to 600 Mc/s)	K. A. MacKinnon
PRA-10	Design Data for Arrays with Reflecting Screens	K. A. MacKinnon
PRA-11	Night Watchman Set No. 2	J. C. Morgan
PRA-12	Design of ASV Long Distance Arrays for Digby Aircraft	K. A. MacKinnon
PRA-13	Design of ASV Homing Arrays for Digby Aircraft	K. A. MacKinnon
PRA-14	An Elementary Introduction to RDF	R. G. Campbell
PRA-15	CSC Equipment Operating Instructions	H. R. Smyth
PRA-16	Radiation Patterns of Paraboloids	J. G. Retallack
PRA-17	Description of CD Receiver Range Gear Box	R. D. Harrison
PRA-18	A Versatile Type of Sweep Circuit	H. A. Ferris
PRA-19	CD Transmitter No. 1—General Description	A. K. Wickson

Report No.	Title	Author
PRA-20	Performance of CD Installation at Duncan Cove	K. A. MacKinnon
PRA-21	CD Transmitter No. 1—Supplementary Information	A. K. Wickson
PRA-22	CD Receiver No. 1—Description and Operating Instructions	H. H. Rugg
PRA-23	SWIC Apparatus Approval Tests	H. R. Smyth
PRA-24	NRC Crystal Testing Unit for 3,000 Mc	J. G. Retallack
PRA-27	Kine Theodolite Tests of APF	W. J. Henderson
PRA-28	Design of a 214 Mc Yagi for CSC	R. E. Bell and C. J. Bridgland
PRA-29	Radiation Patterns of ZPI Collinear for GL Mark IIIC	J. H. Bell
PRA-30	Magnetron Test Equipment—Description and Operation	R. B. Nelson
PRA-31	P'F Equipment	C. W. McLeish
PRA-32	Report on Marine Tests of Cathode Ray Direction Finder (No. 4)	H. R. Smyth
PRA-33	CD Antenna System—Description and Maintenance Instructions	K. A. MacKinnon
PRA-34	Calibration of CD Installation at Duncan Cove	K. A. MacKinnon
PRA-35	Tests of CRDF on Transmission from Aircraft	
PRA-36	Kine Theodolite Tests on APF Set for Br.	W. J. Henderson
PRA-37	Improved Power Supply for Sutton Tube	R. B. Nelson
PRA-38	Milliwattmeter for Measuring Power Output of 10 cm. Oscillators	R. B. Nelson
PRA-39	Crystal Testing—Method No. 2	J. G. Retallack
PRA-40	Description of CRDF for Use on Aircraft	D. W. R. McKinley
PRA-41	Additional Notes to Description of CRDF for Use in Aircraft	D. W. R. McKinley
PRA-42	Specifications and Description on Short Wave CRDF	
PRA-43	CRDF Long Wave for Ship-to-Shore Services	J. T. Henderson, J. W. Bell, and H. R. Smyth
PRA-44	SW-CRDF—Detailed Description	C. W. McLeish
PRA-45	Goniometer Errors—Principles and Practice of Cos $(\omega t - \theta)$	E. L. R. Webb
PRA-46	Control Equipment of CD Installation at Osborne Head	R. S. Rettie and H. H. Rugg
PRA-47	Rotating Dial Sweep Produced by Means of Selsyns	H. A. Ferris
PRA-48	Improved Magnetron Test Equipment—Description and Operation	R. B. Nelson
PRA-49	CD Transmitter No. 2—Description, Operation, Installation and Maintenance	W. C. Wilkinson
PRA-50	Lecture Notes on GL Mark IIIC	R. G. Campbell
PRA-51	CD Receiver No. 2 (Osborne Head)	H. H. Rugg

Report No.	Title	Author
PRA-52	Laboratory Oscillograph Model No. 3	W. B. Johnson
PRA-53	SS2C Equipment	K. C. Mann
PRA-54	Osborne Head CD Displacement Corrector	H. H. Rugg, R. D. Harrison, and G. R. Mounce
PRA-55	Antenna System—Osborne Head	
PRA-57	Trainer for St. Hyacinthe School	H. R. Smyth
PRA-58	Service Manual for Model III Laboratory Oscillograph	W. Johnson and G. R. Mounce
PRA-59	The Theory of IF Amplifiers with Negative Feedback	A. J. Ferguson
PRA-60	The Design of Dipoles for Wave Guide Arrays	K. A. MacKinnon
PRA-61	Tracking a Target in Spherical Coordinates	E. L. R. Webb and K. G. McKay
PRA-62	Unidirectional ASV Beacon Antenna for the R.C.A.F.	T. Pepper
PRA-63	Design of 214 Mc Yagi for Fairmile Motor Launches	R. S. Rettie
PRA-64	Design of Waveguide Feed for Linear Dipole Arrays	W. H. Watson
PRA-65	Design of FAW Antenna	R. E. Bell
PRA-66	Training Device for APF Portion of GL MK IIIC	J. F. Davis
PRA-67	High Speed Oscilloscope, Revised Edition	H. N. Beveridge
PRA-68	Preliminary Reports on Signal-to-Noise Measuring Methods at 10 cm.	G. A. Woonton
PRA-69	Design of a Variable Frequency Transmitter in the Region 75-95 Mc.	J. S. Parsons
PRA-70	Theory of Waveguide Fed Array of Dipoles	W. H. Watson
PRA-71	Some Aspects of the Theory of Coupled (Multiple) RF Transmission Systems	W. H. Watson
PRA-72	Considerations Governing the Feeding of Antenna Arrays by a Single Transmission Line (or Waveguide) with Applications to 70 cm. Variable Frequency VEB	D. C. Brunton and N. Z. Alcock
PRA-73	Use of Elliptically Polarized Radiation for Gap-Filling in the Vertical Pattern of Canadian ZPI Antenna. Part 2	J. H. Bell and E. F. V. Robinson
PRA-74	Operating Instructions for Azimuth Drift Compensation Installed at Osborne Head RDF in February, 1943	K. A. MacKinnon
PRA-75	Coupled Artificial Transmission Lines as Pulse Transformers	J. W. Bell
PRA-76	Omni-directional ASV Beacon Antenna for the R.C.A.F.	T. Pepper
PRA-77	Development of an Adcock Aerial for Use with the CRDF	C. W. McLeish
PRA-78	Portable Precision Wavemeter	K. G. MacKay

Report No.	Title	Author
PRA-79	Delayed Pulse Transmitter for Electrical Alignment of 10 cm. GL Equipment	J. F. Davis
PRA-80	Microwave Linear Array of Slot Radiators	W. H. Watson
PRA-82	P'F Equipment (MK II)	C. W. McLeish
PRA-83	Electrical Pattern Calculator for Linear Arrays	H. LeCaine and N. Z. Alcock
PRA-84	The Coupling of a Resonant Slot to a Waveguide	W. H. Watson and E. W. Guptill
PRA-85	LREW (Long Range Early Warning)	D. W. R. McKinley
PRS-86	PPI for LREW	J. H. Bell
PRA-87	Beam Swing by Phase Changing in a Microwave Array	W. H. Watson and E. W. Guptill
PRA-88	Resonant Slots in Waveguides (second report)	E. W. Guptill
PRA-89	The Design of a Waveguide Array for MEW Using End-fed Dipoles	K. A. MacKinnon
PRA-90	RX/C Description and Operating Instructions	F. R. Park
PRA-91	CDX Radar Test (Devil Battery, Halifax, NS.)	
PRA-92	MEW Antisubmarine—Book A	H. A. Ferris and D. W. R. McKinley
PRA-93	CDX Mark II	R. A. Glaser
PRA-94	Flight Tests of LREW	
PRA-95	Strain Gauge Bridge	S. Korenberg and D. W. R. McKinley
PRA-96	On the Theory of Guide-fed Linear Arrays	W. H. Watson
PRA-97	Preliminary Report on a Combined Searching and Tracking Antenna	
PRA-98	PPI for RX/C	W. M. Cameron
PRA-99	Notes on Matrix Methods in Transmission Lines and Impedance Calculations	W. H. Watson
PRA-100	GL MK IIIC Preliminary Description	J. E. Breeze
PRA-101	Analysis of Errors in CRDF Adcock Aerial	C. W. McLeish
PRA-102	Resonant Slots in Waveguides (third report)	W. H. Watson
PRA-103	Automatic Frequency Control and Frequency Indicators	G. R. Mounce
PRA-104	Longitudinally Polarized Array of Slots	W. H. Watson
PRA-105	Balloon Targets for Centimeter Equipment	G. A. Miller
PRA-106	Theory of Guide-fed Array of Shunt-Coupled Slot Radiators with Mutual Admittances	W. H. Watson
PRA-107	Preliminary Report on Optimum Matching between Crystal and Pre-amplifier without Feedback	
PRA-108	Frequency Characteristics of Slots	W. H. Watson
PRA-109	Further Data on Resonant Slots	J. W. Dodds, E. W. Guptill, and W. H. Watson

Report No.	Title	Author
PRA-111	T/R Cavity and Mixer for 3/4" O.D. 41-ohm Lines	D. Smith
PRA-112	Design of Type E Pulse Network for Use in Blumlien Circuits	W. R. Wilson
PRA-113	A High Speed 9-inch Oscilloscope	E. W. Mazerall
PRA-114	Design of Broad Band Microwave Array of Slots	J. W. Dodds, E. W. Guptill, R. H. Johnston, and W. H. Watson
PRA-116	Development of a Lightweight Pulser for 147 Kw. Output	E. W. Mazerall and W. R. Wilson
PRA-117	Shunt-inclined Slots at 200 Degrees Spacing	J. W. Dodds and W. H. Watson
PRA-119	Rotary Microwave Switch	R. H. Johnston, F. R. Terroux, and W. H. Watson
PRA-120	Guide-fed Arrays of Shunt Coupled Radiators with Mutual Admittance	J. W. Dodds and W. H. Watson
PRA-121	CDX Displacement Convertor (alignment and adjustment)	
PRA-122	Report on ZPI Performance	J. H. Bell
PRA-123	Measurement of Receiver Noise Figures with S-Band Noise Source	A. E. Covington
PRA-124	Preliminary Trials of Type 931	
PRA-125	MZPI Antenna	D. C. Brunton
PRA-126	K-Band Lens Rapid Scanner	R. E. Bell
PRA-127	Power Monitor for MZPI	H. LeCaine
PRA-128	Non-Resonant Arrays	R. E. Bell
PRA-131	Solar Eclipse Observations of the Ionosphere, July 1945	C. W. McLeish
PRA-132	Automatic Ionosphere Recorder	R. Freeman and R. I. Mott
PRA-133	Tests of NRC 200 Mc Distance Indicator at Washington, D.C., January 1946	
PRA-134	Automatic Antenna Pattern Recorder	H. LeCaine and M. Katchky
PRA-135	Transmission Line Fault Locator	W. G. Hoyle
PRB-1	Crystal Units	J. G. Retallack
PRB-2	Notes on Canadian RIS (Mk. II)	H. N. Beveridge
PRB-3	Suggested Method of Coupling from Crystal to First IF Stage in SS2C Equipment	A. J. Ferguson
PRB-4	Visits to MIT, Oct. 17-19 and Nov. 10-Dec. 17, 1942	F. J. Heath
PRB-5	Crystal Units, Supplement to PRB-1	J. G. Retallack
PRB-6	Visit to REL	R. B. Nelson
PRB-7	Long Range Unit Test at Duncan Cove	K. A. MacKinnon
PRB-9	Report of Meeting of IRE, Montreal, U.H.F. Antenna Matching and System	D. C. Brunton
PRB-10	Beacon Tests at Portugese Cove, Jan. 20-24, 1942	H. N. Beveridge

Report No.	Title	Author
PRB-11	Preliminary Report of Performance of MTB Antenna Array Built at NRC from British Specs. H.M. Signal School	G. A. Miller
PRB-12	Visit to Westinghouse, Baltimore, Feb. 3, 1942	J. T. Henderson
PRB-13	Discussion with Bell Telephone Labs Engineers at MIT, Feb. 4, 1942	J. T. Henderson
PRB-14	Visit to MIT, Feb. 4, 1942	J. T. Henderson
PRB-15	Report on Sutton Tube Tests	R. B. Nelson
PRB-16	Motor Torpedo Boat Antennas	H. R. Smyth
PRB-17	Visit to Toronto to Co-ordinate Transfer of Technical Information between NRC and REL	R. E. Freeman
PRB-18	Trial of Exposure Meters	
PRB-19	Visit to Bell Laboratories, Mar. 9, 1942	R. B. Nelson
PRB-20	Progress Report on Variable VEB Antenna Design	D. C. Brunton
PRB-21	Drop Table at Naval Research Laboratories	J. T. Henderson
PRB-22	F.C. and F.D. Sets	J. T. Henderson
PRB-23	Visit to MIT and Portsmouth, April 24-25, 1942	J. T. Henderson
PRB-24	Trip to NRL Anacostia, March 30-31, 1942	J. T. Henderson
PRB-25	Visit to Bell Labs, Whippany, April 1, 1942	
PRB-26	Test of Birmingham U. Magnetron	R. B. Nelson
PRB-27	Magslip Pointers, Mk. II	E. L. R. Webb
PRB-28	Operation of Magslip Receivers from Selsyn Transmitters	E. L. R. Webb
PRB-29	Trip to REL, April 22-25, 1942	J. H. Bell
PRB-30	Trip to REL, April 22-25, 1942	H. A. Ferris
PRB-31	Proposed P.2 Receiver Antenna	T. Pepper and G. A. Miller
PRB-32	Electronic Coordinate Converter and Displacement Corrector	A. W. Y. Des Brisay
PRB-33	Visit to MIT, April 22-24, 1942	
PRB-34	Test of Type E 1534 Magnetron	R. B. Nelson
PRB-35	IFF Test at Toronto, May 1942	F.R. Park
PRB-36	Report on Preliminary Tests of ZPI in Canada	J. W. Bell
PRB-37	IFF as Applied to English Model MK IIIC	I. L. Newton
PRB-38	Six Foot Paraboloid Investigation	C. M. Miller
PRB-39	Tests on SS2C Equipment at Portugese Cove, May-June 1942	A. J. Ferguson and K. C. Mann
PRB-40	Report on Inspection at Montreal, June 26, 1942 of MTB Antenna Array Installed on a Fairmile	G. A. Miller
PRB-41	3 cm. MTB Set: Progress Report No. 1	F. H. Sanders and H. R. Smyth
PRB-42	Notes on Vickers Predictor	J. E. Breeze
PRB-43	Recording and Calibration of Results from CL Angle Test	J. E. Breeze

Report No.	Title	Author
PRB-44	Displacement Correction Graphs	J. E. Breeze
PRB-45	Notes on Sperry Predictor	J. E. Breeze
PRB-46	Tests of Equipment on Corvette "Port Arthur," Halifax	F. R. Park
PRB-47	Visit to G.E.C. and R.L. June 23-4, 1942	E. L. R. Webb and W. Happe
PRB-48	Some Notes on Aided Laying	J. E. Breeze
PRB-49	Visit to Radiation Lab. June 22-24, 1942	R. B. Nelson
PRB-50	Summary of Tests on Rotating Coupler	D. Smith
PRB-51	Trainers for GL MK III	J. E. Breeze
PRB-52	1/10 Microsecond Pulse Ideas	J. E. Breeze
PRB-53	Notes on BTH Crystals	F. H. Sanders
PRB-54	ZPI Interference Tests at REL July 29, 1942	A. K. Wickson
PRB-55	Report on Meeting of Microwave Committee, Washington, July 24, 1942	F. H. Sanders
PRB-56	Proposed Antenna for Early Warning in Forward Areas	A. K. Wickson and G. A. Miller
PRB-57	3 cm. MTB Set, Progress Report No. 2	F. H. Sanders and H. R. Smyth
PRB-58	ZPI Interference Test Using a Controller Aircraft	A. K. Wickson
PRB-59	Operational Aspect of Afterglow on RDF Applications	A. K. Wickson
PRB-60	Summary of Data on IFF MK. III Applications	A. K. Wickson
PRB-61	MTB 3 cm. Progress Report No. 3	
PRB-62	A 10 cm. Mixer for Direction Coupling into a Resonant Cavity T/R Switch	P. W. Nasmyth
PRB-63	Report of a Visit to Radiation Lab and PT Boat Base at Newport, R.I., Sept. 1942	W. L. Haney and H. D. Smyth
PRB-64	A Simplified Video Amplifier	H. N. Beveridge
PRB-65	Report on Microwave Committee Meeting, Sept. 25, 1942	D. W. R. McKinley
PRB-66	Visit to MIT Sept. 16-18, 1942	F. H. Sanders
PRB-67	Trip to REL Oct. 1942, re IFF Mk. III	A. K. Wickson
PRB-68	Report on Trip to MIT Oct. 20-24, 1942	A. J. Ferguson
PRB-69	Survey of IFF Mark III	A. K. Wickson
PRB-70	Calibration of General Radio Wavemeters Type 758A, Oct. 10, 1942	W. H. Watson and E. W. Guptill
PRB-71	Preliminary Tests of CDX Installation at Osborne Head Oct. 1942	J. E. Breeze and H. H. Rugg
PRB-72	Factors Affecting the Design of Display Units for IFF Mk. III	A. K. Wickson
PRB-73	Visit to McGill University, Nov. 3, 1942	J. G. Retallack
PRB-74	MTB 3 cm. Set—Coast and Shipborne Trials, Sept. and Oct. 1942	
PRB-75	Dielectric Measurements	K. G. McKay
PRB-76	Visit to MIT Nov. 30-Dece. 3, 1942	C. H. Miller
PRB-77	Report of Beacon Performance at Halifax, Dec. 1942	R. S. Rettie

Report No.	Title	Author
PRB-78	Report on Tests of Rotatable Yagi for Fairmile Motor Launch Equipment with SW3C	R. S. Rettie
PRB-79	Report on Visit to RDF Group at Esquimalt, B.C. Oct. 1942	H. D. Smith
PRB-80	Comparison of Horizontally and Elliptically Polarized ZPI Antenna	J. H. Bell and E. F. V. Robinson
PRB-81	Automatic Reset Time Delay Relay	C. Fuller
PRB-82	Motor Follower for Crash Boat	W. M. Cameron
PRB-83	Description of FAW Set	
PRB-84	Three Centimeter MTB Set, Progress Report No. 5	
PRB-86	Summary of the Position of GL Mark IIIC as Regards Range Performance	J. W. Bell
PRB-87	Visit to Washington and Bridgeport, Conn., Jan. 14-21, 1943	F. H. Sanders
PRB-88	Radar Committee Meeting at Carnegie Institute, Jan. 15, 1943	H. R. Smyth
PRB-89	A Dipole to Produce Circularly Polarized Radiation for the APF Trainer	W. J. Henderson
PRB-90	Toroidal Cavities for 3 cm. T/R Switches	Dr. Cave
PRB-91	Visit to MIT, Feb. 2-6, 1943	J. G. Retallack, D. W. R. McKinley, and H. A. Ferris
PRB-92	Meeting of Microwave Committee in Washington, D.C., Feb. 19, 1943	W. J. Henderson
PRB-93	Report on VEB Antenna for MEW	W. H. Watson, E. W. Guptill, and F. R. Terroux
PRB-94	Installation and Lining-up Procedure for REB Equipment	W. M. Sharpless
PRB-95	Notes on Tentative Remote Control Design	W. M. Cameron
PRB-96	Trip to REL March 24, 1943, re IFF	A. K. Wickson
PRB-98	Visits to MIT, Raytheon, Sylvania and NRL, April 1943	F. H. Sanders
PRB-99	Discriminator and Band Pass Filter	J. E. Whealy
PRB-100	Preliminary Report on RX/F (MTB 3 cm.)	
PRB-101	Provisional Technical Description of IFF for GL MK IIIC	
PRB-102	Calibration of Analyser Type 2973	C. F. Pattenson
PRB-103	Visits to Bell Labs, MIT and NRL, May 3-17, 1943	F. H. Sanders
PRB-104	Notes on Design of Pulser, X-band	W. R. Wilson
PRB-105	Preliminary Report on Uniform Pulse Lines	E. W. Mazerall
PRB-106	Visits to MIT, Bell Telephone Labs, Holmdel, June 7-11, 1943	W. J. Henderson and R. A. Chipman
PRB-107	An Attempt to Find a Substitute for the 812 Regulator Tube as Used in Sutton Tube Power Supplies	G. W. Hudson

Report No.	Title	Author
PRB-108	Meeting of the Radar Committee in Washington, June 17, 1943	F. H. Sanders
PRB-109	Visit to University of Western Ontario, June 15-27, 1943	H. E. Duckworth
PRB-110	IFF Equipment for GL MK IIIC	W. Happe
PRB-111	Visits to Sperry Gyroscope Co., RCA, NRL and Camp Evans Signal Labs, August 1943	F. H. Sanders
PRB-112	Visits to Westinghouse, General Electric, and MIT, July 23-Aug. 3, 1943	F. H. Sanders
PRB-113	Meeting of Committee on Radar, Washington, Aug. 19, 1943	W. J. Henderson
PRB-114	Visits ot Admiralty Signal Establishment Extension, Bristol, Birmingham University and GEC, Jan. 1944	J. G. Retallack
PRB-115	Limited Experiments with Short Pulses	J. W. Bell
PRB-117	Report on Electrical Scanning by McGill Group	
PRB-118	Salt Water Corrosion of Aluminum Alloys in the Presence of Other Metals	
PRB-119	Modification of ZPI Antenna for 145-155 Mc/s	J. H. Bell
PRB-120	Trip to Victoria, B.C.	E. F. V. Robinson
PRB-122	3 cm. Experimental Set—Preliminary Report	
PRB-123	Microwave Slotted Yagi	W. H. Watson
PRB-124	Thirty Mc. Pulse Signal Generator	
PRB-125	Mutual Admittance between Shunt-Inclined Slots	J. W. Dodds and W. H. Watson
PRB-126	A Transmission Tester for 29-D T-R Tubes	Durnford Smith
PRB-127	The J Antenna	
PRB-128	CNJ	W. Wilkinson and R. S. Rettie
PRB-129	Conductance and Resonant Lengths of Covered Shunt-Displaced Slots	J. W. Dodds and W. H. Watson
PRB-130	Radio Interference Measuring Sets	
PRB-131	Brief Notes on a Visit to Australia, New Guinea, Ceylon and India, Dec. 1943-Mar. 1944	D. W. R. McKinley
PRB-132	Waveguide Couplers for 3BX Magnetrons	A. Covington
PRB-133	Service Notes on RX/C Scanner	W. M. Cameron
PRB-134	Pulse Length Discrimination	B. G. Doutre
PRB-135	A Variable Frequency Sweep with Automatic Variable Slope	W. M. Cameron
PRB-136	Low Level Transmission Tester for Type 64 Tube	Durnford Smith
PRB-137	Report on Oil Seal Tests	J. C. Barnes
PRB-138	Photographs of Magnetron Spectra	J. G. Retallack
PRB-139	Construction of Oxford Type K-Band Local Oscillator, NRC-1	Dr. Cave
PRB-140	Some Data on the Calibrating and Machining of X-Band Thermistors	D. Anderson

Report No.	Title	Author
PRB-141	Tests on 3J31 Magnetrons	W. Barry Clark
PRB-142	Trip to Radiation Lab. MIT, Bell Telephone Labs and Naval Research Labs, July 30 to August 7, 1945	P. D. P. Smith
PRB-144	Application of Auto-Following to Radar AA No. 3 Mk I (GL IIIC)	
PRB-145	Interim Report on S-Band Propagation Trials for CNR-CPR	
PRB-146	Report on Some of the Papers Read at the 1946 Winter Technical Meeting of the IRE	J. E. Breeze and H. LeCaine
PRB-147	Ten Centimeter Propagation Measurements, June 1946	H. LeCaine
PRB-148	Broadcast Receiver Tests—1946 Models	C. F. Pattenson
PRB-149	Report on Physics Congress and Annual Meeting of Canadian Association of Professional Physicists, June 1946	H. G. Byers
PRB-150	Some Possible Arrangements for Operation of MZPI in Conjunction with a Remote Command Post	K. G. McKay
PRB-151	Preliminary Report on a Pulsed Microvoltmeter used in pH Measurement and Control	J. F. Davis
PRB-152	Dielectric Materials with High Constants	

INDEX

Zone position indicator, 71-82 *pas-
sim*; microwave, 82-84